INSIDE
ELECTRONICS

*The How and Why
of Radio, TV, Stereo and Hi-Fi*

by MONROE UPTON

D1494455

A SIGNET SCIENCE LIBRARY BOOK

Published by THE NEW AMERICAN LIBRARY, New York and Toronto
THE NEW ENGLISH LIBRARY LIMITED, London

SIGNET TRADEMARK REG. U.S. PAT. OFF. AND FOREIGN COUNTRIES
REGISTERED TRADEMARK—MARCA REGISTRADA
HECHO EN CHICAGO, U.S.A.

SIGNET SCIENCE LIBRARY BOOKS are published
in the United States by The New American Library, Inc.
1301 Avenue of the Americas, New York, New York 10019,
in Canada by The New American Library of Canada Limited,
295 King Street East, Toronto 2, Ontario,
in the United Kingdom by The New English Library Limited,
Barnard's Inn, Holborn, London, E.C. 1, England

PRINTED IN THE UNITED STATES OF AMERICA

To VICTORIA AUGUSTA

Contents

Contents

Foreword

SOME years ago I wrote a book called *Electronics for Everyone*. I was knee-deep in a new literary venture, *Nuclear Physics for the Underprivileged*, when reports began pouring in that something had gone wrong: *everyone* wasn't buying the book! The title wasn't being taken literally. It seemed to be gravitating mostly into the hands of students, or those with some kind of a direct interest in the subject.

I have therefore postponed my obligations to the underprivileged to write *Inside Electronics*. Here I take aim, not at *everyone*, but at that very large audience of highly intelligent Americans (hear, hear!) for whom my first book on the subject may have been just a hair too technical in parts. This one can serve as an introduction to the first, even as a sort of a review of it, or it can stand alone on its own three feet, which are radio, TV, and hi-fi.

Without too much technical detail, I have tried to lay bare the bones of modern electronic gadgetry, right down to the fundamental diagrams used by the technicians. I believe the time has come when this direct method makes for both a better and an easier understanding than the usual one of the "made easy" books, paraphrasing the original with Van Allen Belt explanations. In other words, the concepts of the technicians themselves should be far less puzzling to the layman if explained in simple, jargonless language.

Electronics covers so much territory today that books on its numerous fields tumble from the presses like children from school at recess time. Digital and analogue computers, network synthesis, induction heating, cryogenics, ultrasonics, etc., etc., all have their specialists. But the theory and circuits we find in radio, TV, and hi-fi are to a large degree basic to all fields. And since these three devices are familiar to all of us, I have concentrated upon them. Using an informal, historical approach, I have tried to show how these electronic devices, as well as a number of others, were evolved from simple, basic theories through the efforts of scientist, inventor, and engineer.

Why does turning up your radio or hi-fi too high increase distortion? What is the difference between the Contrast and Brilliance controls on your TV? Just how does the ultra-linear circuit improve the fidelity of your hi-fi amplifier? For most readers a perusal of this book should remove these questions, together with dozens of others, from the nebulous realm of mystery.

Following an early groundwork of explanation, the later chapters attack the very circuits themselves. This is especially true of the final chapters on high fidelity, which are designed primarily for the layman who assembles his own hi-fi rig. They take him a long step further into the theory of sound reproduction than do most elementary books on the subject. I mean by this that he should acquire such a basic understanding of hi-fi circuits that when he lightly tosses off such phrases as watts of output, impedance match, intermodulation distortion, negative feedback, phase inverter, and damping factor, his conscience will be clear. He'll know what he's talking about.

M. U.

Tucson,
October, 1963

1

THE IMPONDERABLE FLUID

REMEMBER those exciting days of the flying saucers? Remember the stories about strange little men in antenna-spiked helmets who landed their whirling vehicles at some remote spot on our planet, and then took off again with no more than a single, astonished witness to their visit?

Today, all that business of furtive landings and quick departures by interplanetary visitors promises to be a thing of the past. In a cosy little valley in West Virginia, the National Radio Astronomy Observatory has set up an 85-foot radio-telescope dish aimed at two carefully chosen stars in outer space. The scientists are now patiently awaiting signals from "people" on the planets nourished by these stars. Any planet will do. Apparently the scientists are determined that next time, the visits will not be secret. The little men can send us a radiogram before they come, allowing us plenty of time to organize a proper ticker-tape parade up Broadway. This will take care of those mule-heads who scoffed at the eyewitness accounts of earlier landings.

The scientists feel that some of our galaxial neighbors may be much further advanced in science, including the electronic art, than ourselves. After all, we've only known about electricity for about 2500 years. And it was only 360 years ago that we began the experiments which have paid off in such items as the car battery, the image orthicon, and inertial guidance.

A much older and still more potent force in men's lives was largely responsible for electricity's discovery around 600 B.C. Ever since the Bronze Age—perhaps even before— a beautiful, yellowish-brown, translucent substance, which

11

we call amber* has been an article of commerce. Specimens of this fossilized resin from an extinct variety of pine tree have been found in so many ancient tombs that we can draw maps of prehistoric amber routes stretching from the Baltic to the Mediterranean. The men of early Greece had it fashioned into articles of adornment for their female favorites. Thus it was love, primarily (although amber was also used for amulets), that brought the precious stuff thousands of miles across desolate plains and through lonely forests to Greece—where it was found to possess still another virtue besides bringing luck or making a lady more alluring. After rubbing it briskly on chiton or sandal, it miraculously "allured" bits of chaff or papyrus.

The Greeks attributed amber's strange power of attraction to a *soul*. They also philosophized about the "possessive" power of the natural magnet or *lodestone*. Aristotle writes about Thales, Father of Science, who lived three centuries before him, "Thales said that the magnet has a soul in it because it moves the iron."

The Greeks, who liked to simplify problems, dropped the flame of a candle in this same basket with electricity and magnetism. But the modern experimenter saw electricity and magnetism as completely distinct phenomena until the Dane, Hans Christian Oersted, demonstrated their affinity in 1820.

The next step toward today's electronic age wasn't taken until some 2200 years after Thales. In 1600, William Gilbert, an amateur experimenter who also looked after the health of Queen Elizabeth, published a book, *On the Magnet, Magnetic Bodies, and the Great Magnet the Earth*, in which he listed more than 20 other substances besides amber that could, in his words, "take hold of bodies and embrace them, as if with arms extended."

Experimenter Gilbert tried to supplant the Greek anthropomorphic theory of electricity with the concept of an extremely fine, invisible, odorless, even weightless "fluid." When rubbed, this imponderable fluid permeated certain substances to furnish the "embracing arms." It was variously known as an ethereal essence, vapor, or humor; and unctuous stream, an *effluvium*.

The doctor built a device for determining the relative electrical virtues of his rubbed materials. Called a *versorium*, it was the world's first electrical measuring device. A needle was pivoted so that it would turn whenever an electrified

* From the Arabic; the Greek word was *elektron*.

body came close to either end. Any of Gilbert's so-called *electrics* would attract the needle. These electrics included glass, sulphur, sealing wax, all resins, rock crystal, alum, mica, rock salt, and various gems. Here his electrical experiments ended, for the book's main subject was magnetism, which, he mistakenly believed, was something entirely different. Gilbert was the first to explain why the compass needle points north, the reason being implicit in the title: *On the Magnet, Magnetic Bodies, and the Great Magnet the Earth.* Henceforth, for the more cultivated members of society, the compass was robbed of its magic.

The fact that the first electricity was generated by rubbing together two *insulators,* called "electrics," resulted in much fruitless experimentation. *Conductors* of electricity, long ignored, were called "non-electrics."

Pass a plastic comb through your dry hair, and you will charge both comb and hair. The charges "stay put" at the points of contact because electricity doesn't move through an insulator to any great extent, though some charges may pass along its surface, and also escape into the atmosphere—especially if the humidity is high. Use a metal comb, and the charge will leak off from hand to body to earth as fast as you rub. It wasn't until 1737 that a young Frenchman, Charles Du Fay, about whom we shall hear more later, proved that the "non-electrics," such as metals, could also be charged by friction, provided they were properly insulated from the earth by an "electric." When you use the charged comb to attract a tiny bit of paper, you are repeating Gilbert's experiment with his versorium.

After Gilbert, the next natural philosopher to place a rung on the electronic ladder was the mayor of Magdeburg, Germany, Otto Von Guericke. As anyone who has ever successfully fought off sleep through a course of lectures in high school classical physics knows, Von Guericke achieved immortality by demonstrating the power of the atmosphere. After he had emptied a pair of hemispheres (fitted tightly together) with his newly invented air pump, all the King's horses (sixteen of them) couldn't pull them apart. The great man was interested in electricity as well as vacuums and atmospheric pressures. In 1672, he mounted a ball of sulphur "about the size of the head of an infant" on a shaft so it could be turned by a crank like a chicken on a spit. On the supporting base of this machine, he placed "all sorts of little fragments, like leaves of gold, silver, and paper shavings."

He explains one of his experiments as follows: "Stroke the globe with the dry palm, so that it may be rubbed or submitted to friction thus twice or thrice. Then it will attract the fragments, and when turned on the axis will take them along with itself. In this manner is placed before the eye the terrestrial globe, as it were, which by attracting all animals and other things which are on its surface, sustains them and takes them around with itself in its diurnal motion in 24 hours."

The apple with gravity written on it hadn't yet bounced off Sir Isaac Newton's head, so we shouldn't laugh at the Burgomaster for crediting electricity with keeping our mortal coils pinned to this cool little planet till Judgment Day —especially since his experiments in electricity went much farther. He was astonished to learn that the sulphur ball would *repel* as well as *attract*. After a feather had touched the charged ball, it was repelled by it . . . later it was again attracted. He also learned that the feather could be electrified *merely by holding it close to the charged sulphur.*

It was a good century later before a theory appeared that helped to dispel some of the mystery surrounding these puzzlers. This was Du Fay's and Franklin's theory of two *kinds* of electricities: electricities that are alike repel each other; unlike electricities attract. When Von Guericke touched the feather to the sulphur, it acquired some of the sulphur's charge, resulting in repulsion between them. When the feather touched some other object in the room it would be robbed of its charge, and again be attracted to the sulphur. There is always an attraction between either charge, positive or negative, and an uncharged body, as in the case of the plastic comb and the bit of paper. Why this is so, and why the feather was charged when merely held close to the sulphur ball, were questions that had to wait on the electron story for a really satisfying explanation.

The evolution of the Mayor of Magdeburg's frictional electrical machine took almost a century. Around 1705, Newton replaced the sulphur globe with one made of glass. His countryman, Hawksbee, made extensive use of a glass globe frictional generator; he also observed that after a piece of brass foil had touched the charged glass, it was repelled by it, as no report of the Von Guericke experiments had yet reached England. Hawksbee must have held the brass foil by an insulating (electric) handle without realizing the importance of this action. J. H. Winkler of Leipzig, Germany, freed the experimenter's hands by substituting leather pads for pressing against the spinning globe. A better contact with the glass was achieved in 1751 by John Canton, an Englishman, who coated the leather pads with amalgam, an alloy of one or more metals and mercury.

With their hands freed, scientists' imaginations soared, and a number of electricians began charging human beings. The subject was either suspended by silk cords, or placed on an insulating block, his feet in contact with the turning globe. Sparks could then be drawn from his body. (You can make the same experiment by scuffling over a rug when the humidity is low, then reaching for a metal door knob; the spark occurs the instant before you touch the knob.) At the University of Leipzig, G. M. Bozeman hid the charging apparatus behind a curtain and enlisted the support of a pretty girl. He then invited a male in the audience to step forward for a kiss.

> *Ah love's painful disillusion*
> *When fission takes the place of fusion.*

Next, a glass disc, which lent itself to more efficient rubbing, replaced the globe. A glass disc machine (Fig. 2; page 28) was built around 1767 by Jesse Ramsden, an Englishman. The later machines boasted a feature derived from Von Guericke's observation that an object can be electrified merely by placing it in close proximity to a charged object. The phenomenon was called *influence;* we now refer to it as *induction,* or *electrostatic induction.* The charge is said to *induce* a charge in surrounding objects. The induced charge is an opposite (unlike) charge, and this accounts for the attraction. All this can best be visualized through the electron theory, the subject of the next chapter.

The *collector* of the charge on the friction generator didn't have to be in *direct contact* with the rubbed disc, but merely *close* to the disc. Since the experimenter wanted to see sparks, the collector evolved into the shape of a comb, with the sharp teeth pointing at the disc, for it had become common knowledge that electricity will jump much more readily to or from a pointed conductor rather than a blunt one. This was one of the considerations that led Benjamin Franklin to the invention of his spear-shaped lightning rod.

The friction electric generator, whose perigee was in Von Guericke's ball of sulphur, reached its apogee in the 1780's at Haarlem, Holland, in a giant contraption built by the English instrument-maker, John Cuthbertson. It had not one but two glass discs, each over five feet in diameter and complete with eight rubbing pads. The points of the collecting metal combs were one and a half inches from the surface of the disc. With four men sweating on the big crank, enough voltage could be accumulated, it was said, to attract a linen thread 38 feet away. Sparks between the combs and the disc were heavy enough to melt wire one-fortieth of an inch in diameter. Experiments pointed to copper as the best material for lightning conductors, with lead a poor second. This was the first evaluation of the relative merits of metals as *conductors*.

Experiments in *conducting* this high-voltage frictional electricity had begun early in the eighteenth century. The pioneer in this field among the *virtuosi*, as the amateur experimenters were called, was the Englishman, Stephan Gray (1725). Impaled upon the old insulator "electric" and conductor "non-electric" horns, he tried to conduct the electric charge by means of packthread (hempen twine) suspended from nails. This line having failed him, he correctly inferred that the charge had leaked off to earth. So he suspended the packthread by loops of silk thread, attached to the nails. He figured the silk thread would contain the electricity better because it was *thinner* than the packthread. Actually, silk is a good *insulator*. This line, 800 feet long, attracted a feather at its far end. That it worked this well can be attributed to England's damp weather.

After he had read of Gray's experiments, the Frenchman Charles Du Fay showed that metals, when insulated from the ground, can be electrified by friction as efficiently as any of the "electrics," if not more so. In 1737, Du Fay used a wet thread to set a new world record for transference of the electric charge—1,000 feet.

Six years later, another Frenchman, John Theophulus Desaguilieres, published a book in which he dispensed with the unfortunate connotations of "electric" and "non-electric." "But yet non-electrics receive electricity when you bring them near electrics in which electricity has been excited," he wrote. "In order to know that non-electrics have received the communicated electricity, they must be insulated, that is, they must not be suspended from or supported by any bodies but what are electrics; for if a non-electric be touched by another non-electric, which touches a third, and so all the electricity received from the first goes to the second, etc., and from the second to third, till at last it be lost upon the ground or earth."

What John was trying to say in his trips-around-the-mulberry-bush rhetoric, is that a non-electric (metal) can be charged if it is insulated from the ground by an electric (insulator); and only a non-electric will *conduct* electricity. In 1747, Sir Thomas Watson sent a charge through a "non-electric" *wire*, strung over Westminster Bridge, and the charge returned through the "non-electric" water of the Thames. That was the first *ground return*. As he was uncommonly curious as to how fast the effluvium was moving through the wire and the water of the Thames, Sir Thomas rigged up 12,276 feet of wire, supported by insulation. He concluded that the journey along the wire "took no perceptible time."

Electricity generated by friction is high *voltage* stuff, but there is scarcely any *power* in it. You need more current for power. The seat of your pants in contact with the car seat can generate as much as 5,000 volts in your body; the spark that stings your finger the instant before you touch the door handle can be annoying, but it never has proved fatal. A greater *volume* of electricity together with even a relatively low voltage, would be required for that.

Sir Thomas Watson was fortunate in having available for his experiments a revolutionary new discovery of a few years before, the *Leyden jar condenser*. This jar could be charged up by the friction generator: it *accumulated* a small volume of current. This made his work a great deal easier than if he had had to rely directly on the friction machine.

Watson, Du Fay and Benjamin Franklin all contributed to the theory of two *kinds* of electricity. Du Fay was first. He noted that once two pieces of glass had been electrified by rubbing against the same material, they would repel each other. Yet a piece of amber, rubbed by this same material, would be attracted to the pieces of glass. Du Fay concluded that there had to be two varieties of the imponderable fluid. One he named vitreous, after glass; the other resinous, after resins.

Franklin's experiments were much more spectacular: he used human beings. And whereas Du Fay postulated two separate fluids, Franklin managed very well with only one. He placed one of his subjects on a cake of wax, to insulate him from earth, and charged him up by rubbing him with a glass tube. The rubbing also charged the glass tube, which was then used to charge a second subject, also poised on wax. Now, with both subjects "electricized," when they brought their fingers close together a spark would jump the gap. And a spark would also pass between one of them and a third subject, who would stand on the floor uncharged, and therefore more or less grounded. *But the spark between the two "electricized" subjects on the wax was always much stronger.*

From this experiment, Franklin reasoned that under normal conditions, electricity is distributed equally in all matter, like salt in the ocean. However, if this balance is upset by an accumulation of the fluid at any point, the accumulation constitutes a *plus* or *positive* charge. The deficiency, inevitably left by the accumulation, constitutes a *minus* or *negative* charge. Franklin's terms are still in use.

According to this theory, the first subject had received one type of charge (minus), the second subject the opposite type (plus). And a spark between the two *oppositely*-charged subjects would necessarily be stronger than a spark between either of them and the third, uncharged subject who was grounded. As we shall see in the next chapter, the electron theory supports the plus and minus concept to a remarkable degree.

Franklin's fascination with the "electric fire" (as he sometimes called it) began after the discovery of the above-mentioned Leyden jar by Professor Pieter van Musschoenbroek in Leyden, Holland, late in 1745. Because this jar, bottle, or phial, partly filled with water, *stored* a charge, it proved much more useful and convenient to the experimenters than a direct charge from a friction machine. Eighteenth-century experimenters reasoned that the electricity was condensed in the jar, from which comes our word *condenser*. Today, *capacitor* is preferred by most.

Purely by accident, the first capacitor happened to be a jar of water. The professor was holding the jar in one hand, trying to charge the *water* with the voltage from a glass ball friction machine when, as he described it later, "I suddenly received in my right hand a shock of such violence that my whole body was shaken as if by a lightning stroke . . . in a word, I thought I was done for."

What had happened? The charge had come from a chain, dangling in the water from a gun barrel; the barrel had picked up its charge from another chain, which was touching a spinning glass ball. An assistant turned the crank. The earnest experimenters were hopeful that water might "collect" some of the charge. It did. The water was charged *positively*. But they hadn't counted on the fact that the professor's hand, on the outside of the glass bottle, was also charged *negatively*—by induction. (Remember that any object placed in close proximity to a charge receives from it an opposite charge.)

The glass of the bottle prevented the two charges from cancelling each other out, as is true in any capacitor. But when the professor reached over with his free hand and touched the chain dangling in the water, his body provided a path for the charges to combine, and it was then that he felt he was "done for." We would say now that he simply *shorted* the jar's charge through his body.

Iron filings in a jar were found to work as well as water, and lead foil, pasted to the inside surface of the glass, worked even better. A chain, suspended from the cork, pro vided the contact to the filings or foil. Lead foil became preferable to the hand as the outside conductor. John Canton, another Englishman, made the first flat "jar" when he placed a glass plate between two layers of gold leaf. But it was a long time before a substitute for the glass was used.

Today's electronic circuits could no more function without capacitors than birds could fly without feathers. The insulating material between the conducting plates (called the *dielectric*) may be mica, porcelain, glass, waxed paper, air, a plastic, even a chemical. Radio's variable tuning capacitors utilize the air between the plates.

The flat conducting plates may be piled one on top of another, separated by layers of dielectric, to increase the *capacity* (today known as *capacitance*). Two long strips of foil, with waxed paper or a plastic between, may be rolled up and inserted in a cardboard cylinder. Look under the chassis of your radio or TV set, and you will see many of these small, cylindrical *fixed* capacitors. Radio's *variable* capacitors, used for tuning, have one set of plates attached to the dial, plus a second set of stationary plates. When you tune in a station, you move the first set in and out between the second. Maximum capacitance prevails when the two sets are fully meshed, which means that the area shared between them is greatest.

Such prodigious progress was realized from the new "reservoir" of electricity that Tiberius Cavallo, an Italian experimenter living in England, wrote in a book, published in 1777 that ". . . it seems as if the subject would be soon exhausted, and the Electricians arrive at the end of their researches." Nine years before Tiberius died, Volta's chemical cell arrived on the scene, and it is to be hoped that Tiberius had the wit to turn in his crystal ball for a good wet battery.

But even though experimenters could store their home-made lightning in jars, they failed to find any good use for it, aside from mystifying the public with awesome, crackling sparks. For most purposes today, we demand electricity at relatively low pressures—such as the 6 or 12 volts of a car battery, or the 120–220 volts delivered to home or factory— capable of giving us a volume of current that can't be generated by friction, even with the Leyden jar in which to accumulate it.

However, it occurred to many men that the fluid "condensed" in the jars might be useful for sending messages over greater distances than could be attained by smoke signals, or even jungle drums. Most of them were wise enough to confine their ingenious schemes for an electrostatic telegraph to paper.

Vorshelman de Haer's system used ten wires. At the receiver, each wire was connected to a metallic strip, like the keys of a piano. The shocks felt by the operator's fingers were translated into letters determined by some kind of a code. Twenty keys might have simplified or even eliminated the need for a code, if it had occurred to the inventor to use barefoot operators.

A system described in an Edinburgh magazine for February 17, 1753, prescribed 26 wires. The receiver used 26 metal balls, each hanging from a wire. Beneath the balls were pieces of paper with the letters of the alphabet written on them. When the electrified ball disturbed its piece of paper, the sharp-eyed operator quickly wrote down its letter. The inventor also suggested an alternate system with a 26-bell receiver—the largest bell for A, next largest for B, and so on, down to the smallest bell for Z. The author wrote, "Electric sparks, by breaking on bells of different sizes, will inform the correspondent by the sound what wires have been touched." This system allowed musically talented operators to play little tunes for each other during slack periods.

In 1767, an Italian Jesuit, Joseph Bozolus, suggested that long and short sparks, used in a code, would simplify matters. This was the basic idea of the dots-n-dashes that made the Morse and Vail electromagnetic telegraph (1842) superior to England's Wheatstone telegraph, which had preceded it.

Perhaps the only man actually to build a Leyden jar telegraph was Claude Chappe of France in 1790. Chappe used synchronized clocks at the transmitter and receiver. The face of each clock was divided into equal sections, numbered one to ten. A single hand was used. With the hands always pointing to the same number on both clocks, any desired number could be signalled by means of a simple electrical impulse. This telegraph failed for the same reason that all electrostatic systems were impractical, no matter how ingenious. So much of the high voltage leaked off to ground that there was scarcely any left after passing over a mile or so of line.

Franklin did a lot more for the electrical science than formulate his plus-and-minus theory and charge a Leyden phial with the atmospheric electricity that flowed down a damp kite string, proving that a lightning flash is just an outsize electric spark. The Sage of Philadelphia had more curiosity in his bones than a hound dog in its nose, and among the things that puzzled him was just where in the bottle the electricity was "condensed."

He charged up a jar, then removed, successively, the inner and outer conductors. As the charge was not disturbed, he concluded that it must lodge in the glass. Actually, the charge accumulates in the moisture and foreign material that cling to the surface of the dielectric.

Franklin measured the charges and discovered that the positive charge on one side of the glass was always equal to the negative charge on the opposite side. The two charges *balanced* each other, supporting his theory of plus and minus electricities. But his discovery that the *thinner* the glass, the greater the jar's electrical capacity (capacitance) was a bit puzzling. When he moved the plates closer together, which apparently left less room for the weightless fluid, why did the accumulation increase?

Equally mysterious was the fact that some dielectrics take a larger charge than others. Today we give air a *dielectric constant* of one. All other dielectrics have a higher constant. Suppose, for example, we use a 12-volt battery to charge a two-plate capacitor whose dielectric is air. Current flowing into the capacitor soon raises its voltage to 12 volts also, and charging stops because the capacitor's voltage is opposed to the battery voltage, and 12 volts *versus* 12 volts is a "standoff."

Now we slip a sheet of mica between the plates of our fully-charged air dielectric capacitor. *Voilà!* Charging starts again. And by the time the capacitor's voltage has once more climbed to equal the 12-volt charging voltage, the charge will have increased something like six times. This tells us that the mica has a dielectric constant of six. Newly-developed capacitors used in the micro-miniature circuits designed for the electronic equipment in our satellites and guided missiles have a barium titanate dielectric. Barium titanate has a dielectric constant of 15,000.

Du Fay, Watson, Franklin, and the others went about as far as was possible with frictional electricity, so-called, though, as we have pointed out, contact electricity offers a better explanation. The next important advance toward the electron era was the generation of a low-voltage *volume* flow of electricity through chemical change.

In 1800, at the University of Pavia, Alessandro Volta announced his *Voltaic Pile* and his *Crown of Cups*. Volta's experiments, which culminated in the chemical cell, were guided to a large degree by a running argument with a fellow professor, biologist Luigi Galvani, of the University of Bologna. Quite accidentally, in 1780, Galvani discovered that metals applied to the sciatic nerve of a severed frog's leg caused the muscle to contract. After years of experiment, the Bologna professor came to the conclusion that the twitching was caused by an electricity that was inherent in the frog. He therefore called it *animal electricity*. Volta led the scientists who disagreed with Galvani and his followers, and the controversy was a bitter one.

Volta had been generating charges on copper and zinc discs. Holding the discs by glass handles, he pressed them firmly together. His *electroscope,* a simple charge-detecting instrument, which he invented, revealed that contact between the metals generated a positive charge on the zinc and a negative charge on the copper. In an effort to increase the amount of the charge, he tried a *stack* of metal discs, alternating the copper with the zinc. When this failed to work, he separated the discs with pieces of damp cardboard. Merely to increase the cardboard's *conductivity,* he added an acid or salt to the water.

The instant any moisture appeared between the dissimilar metals, a radical change would take place: *contact* electricity was replaced by *chemical* generation. Although of a very low voltage, the chemical action produced a steady, unidirectional *flow* of the mysterious "fluid." The word "flow" later emerged from the inevitable comparison with the movement of water, and is still in popular use.

The moist flesh of the frog between dissimilar metals had functioned in the same manner as Volta's moistened cardboard to generate electricity. But Galvani clung to his animal electricity throughout his life, and Volta never gave up on his contact theory. Both were wrong.

Putting together a chemical cell is almost as easy as mixing a pitcher of lemonade. Place any pair of dissimilar metals in any solution—acid, base or salt—and the combination will generate a potential of up to 1.5 volts, depending upon the materials used. If your teeth have both gold and silver fillings, you are carrying around a potent generating cell in your mouth. Saliva is a weak alkali solution.

Frederick E. Teal, of Berwyn, Pennsylvania, who has read mouth voltages over a period of years, reports that .8 volts is not unusual. Mr. Teal is fearful that these oral voltages, finding their way to the brain, are the cause of a great deal of nervousness, even mental illness. Voltages generated by the brain are on the order of a few microvolts (millionths of a volt). Consequently, a potential hundreds of thousands of times greater could conceivably affect the brain's workings in some fashion. Mr. Teal has assembled considerable evidence that indicates they do. However, he feels that much more research is in order.

2

THE INFINITESIMAL PARTICLE

THE electricity from Volta's chemical cells was termed
galvanic, after Galvani. This was to distinguish it from the
old electrostatic electricity generated by the friction machine
and stored in the Leyden jars.

In 1821, a German physicist, J. T. Seebeck, discovered
that you could also generate electricity from heat. All you
had to do was to heat the *junction* of a pair of dissimilar
metals. The *thermocouple,* as this arrangement is termed, has
long been used as a temperature-measuring device (the voltage
generated is proportional to the difference in temperature
between the *thermo junction* and the rest of the circuit).

After he built the first *electromagnetic* generator in 1831,
Michael Faraday was influential in revealing that there is
only one electricity, regardless of its source. Charge up a
capacitor with electricity from a battery, or from an electrostatic generator, and it's the same old electricity. However,
we still retain the term "electrostatic" for the charge stored
in a capacitor; it means electricity at rest.

Finally, in 1897, J. J. Thomson, at Cambridge University,
England, clarified the nature of electricity itself with his discovery of the *electron.* The concept of the electron makes it
easy to understand both the experiments described in the
preceding chapter and the electronic devices to be covered
in the chapters that follow.

The planetary atom and its main particles probably are as
familiar to the American teenager today as the back of his
school bus driver's neck. But for those oldsters who were
tardy in breaking the comic book habit and would like to
catch up with Junior, here is a quick rundown.

All matter, whether solid, liquid, or gaseous, is composed

of the same building blocks—atoms. There are 92 different varieties of these atoms found in nature, one for each of the 92 different elements, such as copper, lead, silver, oxygen, hydrogen, carbon, uranium, et cetera. (Almost a dozen more have been put together by nuclear physicists.) Until the discovery of the electron, this atom had been considered indivisible, indestructible, eternal. It was then found to possess a central core or *nucleus* consisting of protons and neutrons.

The neutron, as its name implies, carries no electric charge; the proton has a positive charge. The electrons, negatively charged, journey around the nucleus in elliptical paths, like the planets around the sun. In contrast to the planets, they move at tremendous speeds. The mass of each negative electron is only 1/1824th of the mass of each positive proton. In fact, the electron's mass seems to consist of pure electricity, pure energy. However, the *charges* on proton and electron are always equal. This is the secret of the electrical balance of the normal atom.

What distinguishes any one of the 92 atoms from all the others? Chemically speaking, the number of its protons and electrons. The hydrogen atom's nucleus has only a single proton with a single electron in orbit; Helium's atom, atomic number two, has two protons (plus two neutrons) in its nucleus, with two "whirling" electrons. And so we go on up the scale to the heaviest atom, uranium, with 92 protons and 92 electrons, although its nucleus contains enough neutrons to bring its atomic weight up to 238. However, in understanding electronics, as well as chemistry, we are only concerned with atomic number, from one to 92. We can neglect what goes on in the heart of the atom, whose ultimate secret, despite the prolonged and valiant attacks upon it by the world's nuclear physicists, with their giant particle accelerators and intercepted cosmic rays, has proved almost as elusive as the abominable snow man. Even when all of the nuclear particles (of which more than 30 have been discovered) have been catalogued and their forces understood, the final conquest of the atom will most probably have to wait, for it is by no means certain that the nuclear mystery, which probably holds the key to all of nature's laws, if such *laws* exist, will ever fully yield to mathematical formulae and the particle theory.

Most substances are *compounds* of the 92 natural elements. A compound has in it groups of atoms of the combining elements. Each little group of atoms is called a *molecule*. The molecule of common table salt, for example, has an atom of sodium in chemical combination with an atom of chlorine.

Now consider an individual atom with its equal numbers of protons and electrons, all the charges equal. Electrically, this atom is perfectly balanced. The positive balances the negative, and no electricity can be detected. But take away one or more electrons, and the atom is left preponderantly positive; or, add one or more electrons, and the unbalancing is on the negative side.

This recalls Franklin's theory of the nature of electricity, though that astute amateur conceived of the particles of his electric fire, the imponderable fluid, as separate from matter, not "broken off" from larger particles. (The modern atomic theory, as first formulated by John Dalton, English schoolteacher, wasn't developed until 1803.) And Benjamin also had the positive charge moving toward the negative charge, whereas the opposite is largely true. It is only in the gases, and to a lesser extent in liquids, that the (roughly) two-thousand-times-heavier, positively-charged nuclei are free to move, always in an opposite direction to the movement of the more abundant negative electrons. In a solid, such as the conventional copper wire conductor, *electrons alone comprise the current flow*.

An atom that has become unbalanced, either by gaining or losing electrons, is called an *ion*. This means, of course, that there are both positive and negative ions. The atoms on the surface of a substance can always be unbalanced through *contact* with another substance, best achieved through *friction* between them. (The heat generated by the friction has nothing to do with it.) Rub any two dissimilar substances together, and one of them will always acquire some of the other's electrons, charging it negatively. The other substance, with positive ions equal in number to the lost electrons, will be positively charged.

But how about electrostatic induction, first noted 300 years ago by the Mayor of Magdeburg? Why does an uncharged body acquire a charge of an opposite sign when it is merely brought *near* a charged body?

Suppose we place a positive charge near an uncharged piece of metal. The free electrons in the metal will be attracted by the positive charge, and move as close to it as possible. Thus, the part of the metal nearest the positive charge acquires a negative charge. Result: attraction. And note that since all of the metal except that part adjacent to the charge has lost electrons, that part acquires a positive charge (Fig. 1).

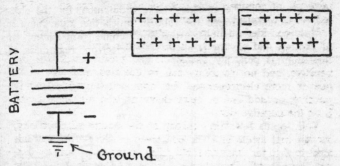

FIG. 1. A positively-charged body attracts the negative charges in an uncharged (neutral) body, charging it as indicated. A negatively-charged body would charge the adjacent body in the opposite way.

On the other hand, if our charge is negative, it repels the electrons in the uncharged metal, leaving the area near it positively charged, and the rest of the metal, with its surplus electrons, negatively charged. These experiments explain why an *induced* charge is always an *opposite* charge, resulting in attraction between them, as in the case of the comb and the bit of paper. They also explain the spark that results if the charge is great enough, as in the case of the comb on the friction machine. Opposite charges are always striving to neutralize each other and thus restore nature's balance. In the case of an insulator, there are always enough electrons on or near the surface to create some kind of a charge.

The Ramsden electrostatic generator of Fig. 2A is shown grounded. This adds greatly to the machine's efficiency. The electron theory makes it clear just why this is true. Friction between amalgam and glass charges the glass positively, negatively charging the amalgam. The amalgam's surplus electrons are absorbed by the earth. The positive glass pulls up the electrons into the points of the comb as close to it as possible, leaving the rest of the comb electron-deficient and therefore positive (Fig. 1). This positive part of the comb attracts fresh electrons from the accommodating earth. When a pressure builds up high enough between positive glass and negative comb, there is a spark; and for the duration of the spark, the electrons travel round and round through the circuit, the same kind of circuit that Watson used across the Thames and back.

We can even draw an analogy between Ramsden's primitive machine and electronics' basic radio tube, for the electrons circulate through the tube in much the same manner (Fig. 2B). The tube's electrons don't come from friction or mother earth, but an electron is an electron, regardless of how we start it on its journey.

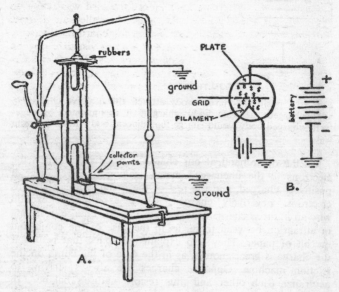

FIG. 2. Electricity flows through a modern radio tube (B) in much the same way as it did through Ramsden's electrostatic generator (A).

Heat is the prime source of electrons in the tube. An incandescent *filament,* similar to any lamp filament, or a surrounding metal sleeve, which the filament heats, is said to "boil" off the electrons. Since the electrons are negative, this electrode in the tube is called the *cathode.* (Any negative electrode, even that of a storage battery, is a cathode.) A steady positive voltage on a metal *plate,* or *anode,* attracts the cathode's electrons. The negative end of the plate voltage is connected to the cathode. Thus, the electrons, attracted to and absorbed by the positive plate, are returned to the cathode through the plate voltage source. This source may be a battery, as shown in the diagram, a direct current generator, or the alternating current from the power company, provided it has been changed to a direct current (that is, *rectified*). (In almost all instances, the tube's plate voltage must be steady and uni-directional, what we term a *direct current*.)

When in 1906, Dr. Lee de Forest placed a grid of wires *between* a hot cathode and a cold plate, all enclosed in an evacuated glass tube, the electronic age was ready to take off; though it was some years later before the bird was really on the wing. De Forest's little tube, which he called an *Audion,* was first used as a detector-amplifier in shipboard receivers.

The electron theory also makes it easy to visualize why an insulator insulates and a conductor conducts. The atom's circling electrons are confined to orbits that are called *shells.* Each shell holds only so many electrons before a new one, beyond it, is required. The shell nearest the nucleus holds two electrons, the second one eight, third 18, the fourth 32.

Things get a little complicated after that. Five gases—helium, neon, argon, krypton, xenon, and radon—are unique in that their outermost shells are completely filled. As a result, they refuse to combine with each other or with other elements. They are said to be *inert*. But the remainder of the 92 elements in the basic periodic table have outer shells that lack their full quota of electrons. Chemistry explains the combination of elements that form compounds on the basis of these vacancies. Atoms tie themselves together by shifting one or more electrons from atom to atom. For example, sodium has only a single electron in its outer shell, while chlorine has seven. So when we bring sodium and chlorine together, the lone sodium electron slips over into the chlorine outer shell, giving it its full quota of eight electrons, and leaving the sodium with its outer shell completely filled also. Then both atoms are stable. And the sodium atom, having lost an electron, has become a positive ion; while the chlorine atom, having gained an electron, has become a negative ion. Thus, the atoms are joined by electrical attraction to form an ion (charged molecule) of the compound sodium chloride, better known as table salt.

The technical word for this is *valence,* specifically *electrovalence,* for there's another kind called *covalence*. With covalence, the outer shell electrons also interlock to form compounds, in which the atoms are grouped in molecules instead of ions. In either case, however, the compounds, such as glass, paper, mica, porcelain, et cetera, cling to their "meshed" outer shell electrons with such tenacity that they can be loosened only by an extremely high voltage. To create a current flow through these materials is so difficult that we can use them for insulators. But most of the metals present an entirely different problem.

Most metals are composed of crystals, as are salt, sugar, quartz, and sulphur. Unlike the vitreous type of solid, such as glass, a crystal has its atoms arranged in a geometric pattern. Strike certain crystals with a hammer, and the break occurs along definite planes.

Outer shell electrons of the atoms aren't always necessary to their formation into this regular pattern or lattice. For example, each of our four best solid conductors—platinum, gold, silver, and copper—has a single electron in its outermost shell with such weak home ties that it is constantly on the move, jumping from atom to atom. Any kind of an electric *generator,* such as a battery, or a wire cutting across a magnetic field, gives these electrons a *directed motion,* a phenomenon we call current flow.

So many of the atoms in our universe are constantly losing or gaining electrons that to say electricity is ubiquitous would be the understatement of the millennium. Ever since Gilbert's discovery in 1600 that electricity was not confined to amber (the Romans had electrified jet by friction), science increasingly has realized the important role electricity plays in the material scheme of things. Most of our physical functions are described now as electro-chemical in nature. These functions are accompanied by a "surge" of electrons along the body's network of nerves. Today, we amplify sufficiently their feeble voltages to provide a record on paper tape of how our heart is acting, our brain operating, our stomach digesting, or one of our muscles contracting.

The gas we call the earth's atmosphere, which extends outward in increasingly rarefied form for hundreds of miles, is fairly "alive" with ions. Electrons are added to or knocked out of the molecules of this gas, mostly by radiation from the sun. Cosmic rays from outer space (mostly protons) also contribute to the ionization. Science has recently determined that breathing negative ions is healthful, while positive ions have the opposite effect upon us. Positive ionization increases with smog and during thunderstorms. A device for generating negative ions in the home is now on the market.

The application of the electron theory to the functioning of the chemical cell starts with this basic fact: in the case of a liquid, no outside source is required for ionization. Add an acid, a salt, or a base to water, and ionization takes place spontaneously! Both positive and negative ions appear. And if you place a pair of dissimilar metals in the solution, each reacts differently.

For example, take a simple experimental cell, similar to one of Volta's originals, consisting of a sulphuric acid solution with copper and zinc electrodes. The sulphuric acid molecule is written H_2SO_4, which simply means it contains two hydrogen atoms, one sulphur atom, and four oxygen atoms. In solution, this molecule breaks up into three parts: two separate hydrogen atoms, and what is left of the molecule, which is SO_4. In the breaking-up process, all the electrons don't stay at home in their atoms. Both hydrogen atoms lose their electrons, which turns them into positive ions. The lost electrons attach themselves to the SO_4 part of the original molecule, making it a double negative ion written SO_4^{-2}. These double negative ions attack the zinc, pulling positive ions from it to charge it negatively. On the other hand, the positive hydrogen ions neutralize electrons in the copper, leaving it positively charged.

Now suppose we connect a circuit, consisting of a coil of copper wire, between the two terminals or the zinc and copper electrodes. Since copper is our best conductor, next to silver, it aways has a great many vagrant electrons "on the road." However, there are no concentrations of these itinerant electrons. They are fairly evenly distributed throughout the copper. So you couldn't say that the copper coil as a unit, or any sizable part of it, was *either* positive or negative. For all practical purposes, it is electrically balanced, or, to use a more technical term, it is neutral. Therefore, the negative (zinc) electrode's surplus electrons will exert a pressure on it, and many of them will move into the wire.

Meanwhile, at the other end of the coil, at virtually the same instant, the positive (copper) electrode's *electron deficiency* will have the welcome sign out for any and all free, travelling negative electrons. The negative electrode pushes, the positive electrode pulls, and the resultant *directed movement* of electrons through the coil provides us with a current flow. The zinc is slowly "eaten up" through its constant loss of positively-charged atoms, which are positive ions. There is no dodging electricity's price tag.

3

THE MARRIAGE OF ELECTRICITY
AND MAGNETISM

ELECTRONICS is a highly-sophisticated offspring of the age of electric power, scion of the marriage of electricity with magnetism. This event marks the highest peak of technological progress between James Watt's addition of a separate condensing chamber to the Newcomen engine in 1765, and the first self-sustaining nuclear chain reaction obtained by Enrico Fermi and his co-workers on the University of Chicago squash court in 1942; though a Gallup poll would certainly reveal that man's daily ration of quiet desperation was nullified to a greater degree by still another technological marriage—nylon with rubber—a union that gave us that vitalizer of male awareness and strengthener of female morale, the three-way-stretch foundation garment. And if progress continues the next unifying step should be free-floating bras with electronic feedback for optimum cleavage.

The electromagnetic nuptials were long overdue. Franklin's experiment, in which he magnetized steel needles with current from a bank of Leyden jars, should have been the tip-off. But instead of paying heed to it, the electricians wasted countless hours trying to find some magnetic emanation from an *electrostatic* charge, and electricity and magnetism continued to be generally regarded as separate and distinct phenomena.

The classic experiment that finally united electricity and magnetism took place in 1819, in a classroom at the University of Copenhagen, under the guiding hand of Professor Hans Christian Oersted. His instruments were a large voltaic battery, a loop of wire, and a small compass. Oersted held the compass over the wire, which carried a heavy current from the battery. Momentarily interrupting the current, he maneuvered the compass until the needle was parallel with the wire, then, as he restored the circuit, he saw the needle swing around until it came to rest at right angles to the wire. Reversing the direction of current flow caused the needle to swing around until it lay across the wire while pointing in the opposite direction. The right angle relationship between the current's *magnetic force,* as indicated by the needle, and the current *direction* was a surprise—for the rule that all forces should act along a straight line, as with the gravitational force, was believed inviolate.

And so it was revealed, three decades after the *discovery* of current flow in the body of a frog by Galvani, that a wire carrying current is surrounded by the same magnetic force that surrounds the natural magnet or lodestone, which had intrigued man for thousands of years. The new, *electromagnetic* force had two profound advantages: it could be much more powerful, and it could be controlled merely by regulating the size of the current flow through a coil.

The first practical application of electromagnetism was to be the telegraph, followed by the telephone. Joseph Henry laid the foundation for these inventions when, in 1831, at the Albany Academy, he magnetized iron at a distance of one mile. The magnetized iron attracted one end of a permanent magnet centered on a pivot, causing it to swing around and strike a small bell. (The bell is still preserved at the Academy.)

Four years later, Henry installed an experimental telegraph system on the Princeton campus between his laboratory and his wife's kitchen. ("Have Appetite Will Travel" preceded Morse's "What hath God wrought" message by nine years.) In conjunction with a single overhead wire, he used a *ground return*, which was no novelty (remember Watson's over the bridge and back through the Thames circuit?) although his electromagnetic *relay* was. The relay, which was to prove indispensable to the electric telegraph, consisted of a very sensitive electromagnet, just powerful enough to make and break a local circuit, whose heavier current operated the receiving device. Thus, all the basic materials for a practical telegraph were waiting for Morse and his associates.

When you switch on your car's self-starter, the brute force that starts the motor churning is the same that moved the compass over the wire in Oersted's epochal experiment. Within a short time after that discovery, André Marie Ampère, the great French physicist, demonstrated the manner in which two current-carrying *wires* react to each other like two magnets. For maximum effect, the two wires had to be parallel; at right angles, the magnetic force between them was zero. Secondly, he found that when the direction of current flow in both wires was the same, the wires were attracted to each other; when the direction of current flow was opposed in the two wires, they repelled each other.

Dominique François Arago (1786-1853), noted French physicist and astronomer, formed the wire into the shape of a coil or *helix* and suspended it horizontally by a thread so that it was free to turn. When he fed it with a current, it aligned itself in a north-south position, as if it were a compass needle. The *coil* functioned as a *bar magnet* does (Fig. 3B).

The human mind, shackled by common sense, has always stumbled over the concept of "action at a distance" without some material connection between two bodies. Even Newton believed in a *medium* for his force of gravitation. It was Michael Faraday, son of a London blacksmith, who furnished the media for the forces of electricity. He called them *fields,* composed of *lines of force.* The electrostatic field surrounds the charge generated by friction, or is stored in a capacitor; the magnetic field surrounds the *charge in motion* that we call a current flow. The current flow also retains the electrostatic field. Without it, we wouldn't have radio and TV, for the waves that bring us sound and pictures are *electromagnetic;* this means that they are a combination of both fields—the electrostatic, usually called the electric field, and the magnetic. (See Fig. 14; page 66.)

Faraday's lines of force are really as imaginary as Thales' soul or Gilbert's effluvium, but they enjoy the relative advantage of being immensely practical. The mathematician rates the intensity of the field by the *number* of these imaginary lines of force per unit area. The earth's magnetism has an intensity of around ½ *gauss.* (The *gauss* is named for Karl Friedrich Gauss, the great German mathematician.) This means half a line of force for each square centimeter. The permanent magnets in the best of today's loudspeakers, made of an alloy of several metals, mostly aluminum, nickel, and cobalt (*alnico*), may have an intensity as high as 17,000 *gauss.* The limit in a magnetic core, such as the one used in the output transformer of a hi-fi set, is around 60,000 *gauss.* Beyond that is *saturation,* when the core's magnetism stays the same no matter how much more current flows in the coil. Laboratory currents of thousands of amperes have produced momentary magnetic intensities in air or vacuum of close to a million *gauss,* powerful enough to disintegrate a piece of tough steel.

The lines of force of Faraday's fields make electromagnetism relatively easy to understand. In our mind's eye, we see the electrostatic lines moving straight out at right angles from the charge, the magnetic lines circling a current-carrying wire (Fig. 3A). In the case of Arago's coil, the circling magnetic lines must all move in one direction down through the coil's center, and in an opposite direction on the outside (Fig. 3B).

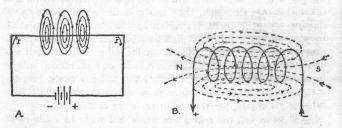

FIG. 3. (A) The magnetic lines of force circle a conductor carrying a current. The electrostatic lines are at right angles to the current flow. Reverse the direction of current flow, and the direction of the magnetic lines also reverses. (B) When the conductor is formed into the shape of a coil, the circling lines are bunched so that they move in one direction, as indicated by the dotted lines. Thus a current-carrying coil has a north and south pole like a bar magnet.

Ampère's pair of wires were attracted to each other when the current flow was in the same direction through both; there was repulsion when the direction of one current was reversed. Now when the currents through two parallel wires are in the same direction their surrounding lines of force move in opposite directions; and when two such currents are opposed their lines of force move in the same direction. We must, therefore, conclude that opposing lines of force attract, while lines in the same direction repel. Keeping this in mind enables us to understand the operation of that small electric motor we call a loudspeaker. (Fig. 58; page 182.)

Attached to the speaker's paper diaphragm is a small coil of a few turns of fine copper or aluminum wire called the *voice coil.* The voice coil fits like a sleeve over one pole of the permanent magnet (alnico); it is suspended so that it is free to slide in and out for a small fraction of an inch. The other pole of the permanent magnet is split so that it almost completely surrounds the voice coil. Now if we trace the relationship between the lines of force of magnet and voice coil we shall see why a rising and falling signal current through the coil moves it in and out.

The permanent magnet's lines of force pass straight through the coil from the center pole to the two outside poles, parallel to its turns. Since the copper or aluminum voice coil is non-magnetic, it is not affected—until current flows through it, making a magnet too. The coil's lines of force encircle each individual turn, moving, in one direction on one side of each turn, and in the opposite direction on the other side. On one side, they move in the same direction as the big magnet's turns (repulsion), while on the other side, they move in the opposite direction (attraction). With attraction on one side and repulsion on the other side of each individual turn, the entire coil moves in one direction, either in or out. And if a current through the coil moves the coil *in,* a reversal of the current's direction, which reverses the directions of its encircling lines, will move the coil *out.*

How does the voice coil act when fed with an *audio frequency* something like the one pictured in Fig. 46 (page 148) which is a combination of frequencies ranging mostly from 50 to 10,000 cycles per second? The little coil will flutter like a mother hen at the moving shadow of a hawk on the ground. Not only will it dart quickly in and out in response to the signal's relatively weak higher frequencies, as the current rises and falls, but when the oscillating current reverses direction at the end of each half cycle of the fundamental frequency, the coil will move to the other side of its center position of rest. To respond accurately to the extremely complex mixture of frequencies from an orchestra, for example, is quite a chore to ask of this electro-mechanical device, but even the cheaper ones make a brave attempt.

Oersted's compass needle proved that electricity, in the form of a current flow, is always accompanied by magnetism —in effect, it *produces* magnetism. Faraday, and the American, Joseph Henry, among others, were hopeful that the process could be reversed and magnetism be made to produce electricity.

An electrostatic (electric) charge, obtained by contact, induces an opposite charge in the nearest portion of any object brought near it (Fig. 1; page 27). Faraday didn't have much more than this to go on when he began his attempts to induce a flow of current. He connected the steady, uni-directional current (d-c) from a battery to a coil, a *primary* coil. Close to it he placed another coil, a *secondary* coil. Although there was no physical connection between them, the secondary coil shared the lines of force of the electromagnetic field that was created in the primary coil by the battery current. In the light of Ampère's experiments, all of the turns in both coils were wound parallel with one another (Fig. 4).

But no amount of current passing through the primary, regardless of the kind of wire used, or the size of current flow, seemed to induce any current in the secondary coil. Finally, after fruitless months of experiments, Faraday was alert enough one day to note that the instant he either made or broke the battery connection, there *was* a transfer of current. But it was only a quick *surge* of current. The instant the connection was made, the compass needle would move to one side of its center position, then flop back again. When the connection was broken, the needle would move to the opposite side of center, then flop back once more. He had missed it before because the needle had returned to zero position so quickly.

What had happened was this: *making* and *breaking* the

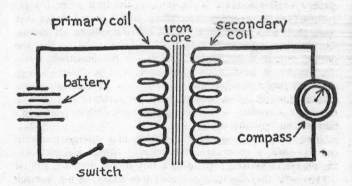

FIG. 4. This circuit illustrates Faraday's experiment, which first demonstrated that electricity could be obtained from magnetism. Only when the current was rising, after the switch had been closed, or falling, after it had been opened, did the compass needle move. The iron core, indicated by the parallel lines, was inserted later; it greatly increased the *amount* of current *induced* in the secondary coil. From this experiment came the spark coil, which was the first wireless transmitter, and later the transformer.

battery's direct current had induced an *alternating* current in the secondary coil.

Ironically, Faraday wasn't particularly pleased with this great discovery. He was seeking in the secondary coil the same steady, uni-directional current that came from the battery. What good was a current that kept jumping up and down and reversing direction every other jump?

It was evident that for any transfer of electricity between two separate coils, the current in the primary coil had to be *changing;* either rising after the contact was made, or falling after the contact was broken. Some relative *movement* was necessary.

The first practical device to emerge from this revolutionary experiment was the spark coil. Both Page in America and Rumhkorff in Germany placed an *interrupter* at one end of an iron core upon which both the primary and secondary coils were wound, one on top of the other (Fig. 11; page 63). The interrupter automatically kept the battery current through the primary continuously *changing.* Half a century later, this interruptor, or vibrator, became what is now the old-fashioned door bell.

The spark coil soon became a valued laboratory device for producing high voltages. The voltage ratio between the primary and secondary is the same as the turn ratio: for example, twice as many secondary turns means twice the voltage; half as many turns means half the voltage. Of course, current flow is proportionately reduced, or raised, so that the power output is unchanged, except for the inevitable losses that occur as heat in coil and core. The spark coil's high voltages made possible the Crookes tube experiments from which came the discovery of both the electron and X-rays. And the Crookes tube eventually evolved into television's camera and picture tubes.

The spark coil was also the world's first wireless transmitter. In 1888, a young research scientist named Heinrich Rudolph Hertz hooked one up to a tiny dipole antenna, similar to the ones that decorate our roof tops today, in his research laboratory at Karlsruhe, Germany. Hertz was testing the theory advanced by James Clerk Maxwell at Cambridge that electrical waves can be generated that are the same as light waves except for their lower frequency. And sure enough, Hertz not only succeeded in generating the waves, he also discovered their velocity to be 300,000,000 meters per second. He noted that the waves were reflected by an electrical conductor, such as a sheet of metal (the basis of radar), while at the same time they passed through materials we know as insulators. It was obvious that the Hertzian waves, as they were long called, moved through space in a straight line as light does. Because of the earth's curvature, this seemed to make them useless for anything but very short range telegraphy without wires. We shall discuss the Hertz experiments in Chapter 5.

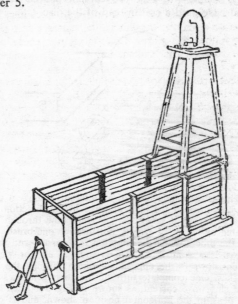

FIG. 5. Faraday's magnet and 12-inch copper disc comprised the world's first electromagnetic generator. The compound bar magnet was made by Dr. Gowin-Knight, of the Royal Society, and can be seen at the Science Museum, London.

Faraday's next fruitful experiment was one that led him to the construction of the world's first electromagnetic generator. He thrust an ordinary bar magnet down through the center of a coil. The magnet's down-thrust moved the compass needle in one direction; the pull-out moved it in the opposite direction. This was just another way of subjecting a coil to the *changing* lines (rise and fall) of a magnetic field.

Faraday's aim was to rig up a machine that would provide a *continuous* movement of a conductor across magnetic lines of force in order to generate a *continuous flow of current* (Fig. 5). He fastened a 12-inch copper disc to a shaft so that when the crank was turned, the disc's outer edge would cut across the lines of force between the north and south poles of a big permanent magnet, the biggest one in England. Two contacts were mounted so that one of them brushed against the shaft, the other one against the edge of the disc. And it worked. The "brushes" picked off a continuous current flow. The only trouble was that the current proved very feeble; he should have used a coil instead of the disc.

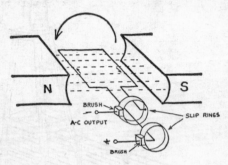

FIG. 6. This is the single loop of an alternative current generator. When it cuts through the lines of force in the direction indicated by the large arrow, it generates a current flow through the loop, as indicated by the small arrows, out through one brush, and back through the other. When the situation is reversed, and the side of the loop that was moving down is moving up (or vice versa), the current flows in the opposite direction through the slip rings and brushes. One complete turn of the loop generates two half cycles of sine wave current, as shown in Fig. 10. Each half cycle is at its peak in the horizontal position shown, and is cut across the lines of force at right angles. At the vertical position, the current generated is zero because, for an instant, the loop is moving in the same direction as the lines of force and not cutting across them.

Nevertheless, all of today's electromagnetic generators, from radio's tiny microphone to the giant alternator at the foot of a dam, stem from this crude contraption built by the self-taught son of a London blacksmith. The list of inventors who contributed to its evolution reads like the roll-call at an international scientific congress: H. Pixii, Ernst Warner von Siemens, Antonio Pacinotti, Zenobe Theophile Gramme, Henry Wilde, Sir Charles Wheatstone, John Hopkinson, George Westinghouse, Nikola Tesla, William Stanley, and others.

A coil, rotated between the two poles of a magnet, generates an alternating current because after each half revolution the coil's turns change their direction through the magnetic lines of force. And as Fig. 6 reveals, this reverses the direction of the generated current.

To obtain a direct current (d-c) from a machine that wants nothing more than to generate an alternating current (a-c), it was necessary to "turn around" every other half cycle of current. This was accomplished by means of a *commutator* (Fig. 7). This device consists of a number of metal

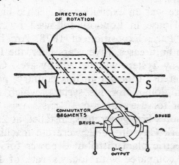

FIG. 7. Single loop of a direct current generator: When you connect each loop to a pair of segments on a commutator, every other half cycle of generated voltage is turned around, producing a direct current (d-c).

segments, separated by insulation. The two terminals of each loop are connected to each pair of segments. As the shaft turns, the brushes make contact with a different pair of segments, one after another, automatically reversing the direction of every other half cycle of generated voltage. Thus current always flows out of the machine in the same direction, fulfilling Faraday's dream of a direct current generator. The numerous loops together comprise what is called the *armature*.

Your car's generator is one of these d-c machines with a segmented commutator because a direct current is necessary to charge the storage battery. But for industrial and home use, a return to the machine's inherent a-c nature proved necessary.

There were many reasons for this. Sparking at the commutator when generating a heavy current was one problem. In raising and lowering voltages with heavy currents, the spark coil's interruptor was also useless for this same reason;

whereas, with a-c, a *transformer* with no interruptor can be used.

Today's long transmission lines demand extremely high voltages to keep down the losses in the wires; some of these voltages are close to half a million. Alternating current generators are even going into some of Chrysler's late model cars, together with a rectifier to change the a-c to d-c for the battery. There has been a vast improvement in small rectifiers in late years.

In 1867, there was an exhibition of electric lighting on top of Burlington House in London, followed by a similar exhibition at the Paris Exposition of 1878. Arc lighting from a-c was used in both cases. Five years later, the London Metropolitan Railway System was lighted by arcs. George Westinghouse pioneered a-c power in this country, paying Nicola Tesla a million dollars for his system, which included the induction motor for use with a-c.

Thomas Edison's first generators, installed at 255-257 Pearl Street, New York City, in 1882, were d-c machines. This pioneer steam-electric plant distributed power for 24 cents per kilowatt hour, compared with today's rate of from one to five cents. (A kilowatt hour is equal to 1,000 watts used for one hour, or, in other words, ten 100-watt lamps burning for an hour.)

Many microphones and *phonograph pickups* are electromagnetic generators with coil and permanent magnet. But instead of *turning* in the magnetic field, the coil is made to *vibrate*. Moving one way across the lines of force generates a voltage in the coil in one direction; reversing the coil's direction reverses the direction of the voltage. So an *alternating* current is still generated. The mike's coil is attached to the diaphragm, which is moved back and forth by the air waves of sound. In the phonograph pickup, the motive power is furnished by the needle vibrating in the grooves of the record.

A contemporary of Oersted, Ampère, Faraday, and Henry, who also contributed a cornerstone to today's vast electronic edifice, was a German schoolteacher by the name of Georg S. Ohm. Ohm's contribution was his *Mathematical Theory of the Galvanic Circuit,* published in 1827, five years before Faraday's generator. We now refer to it simply as *Ohm's law.*

Ohm saw a possible analogy between the conduction of heat and electricity. Some metals offer less opposition to the transfer of heat than others. Place a stainless steel spoon and one made of silver in a hot cup of coffee, and the heat will climb to the top of the silver spoon first. The superior heat conductors, copper, silver, gold, and platinum, are also the best conductors of electricity, strangely enough.

Of course, we can't carry this analogy too far, because in practical applications, electrical conduction is seldom a matter of travel time. Connect a circuit to a battery, and no matter what the length, travel time for an electron movement between its terminals, from negative to positive, will be almost nothing flat—the speed of electricity in a conductor is close to 186,000 miles per second. In electronics, we are mainly interested in how long it takes a current to rise to its peak and then fall away to zero. This *frequency* can be controlled by means of capacitors, coils, and resistors, a matter we shall explore in the next chapter.

Ohm's law has since enabled us to think of electricity as possessing both *pressure* and *rate of flow*, with the circuit offering a definite amount of opposition to the pressure. To substitute the technical expressions, an electric circuit has a voltage (pressure), a current (rate of flow), and a resistance (opposition). The rate of current flow is directly proportional to the voltage, and inversely proportional to the resistance.

The units of voltage, current, and resistance acquired names at the Electrical Congress in Paris, in 1881: the volt, after Volta; the ampere, after Ampère, and the ohm, after Ohm. Since then, every electrical worker, no matter how humble, has had to be on familiar terms with the relationships between these terms. $I = E/R$, the I for current flow in amperes (think of intensity), the E for the electromotive force in volts, and the R for resistance in ohms. And if $I = E/R$, then $E = IR$, and $R = E/I$.

In considering a problem, we had best start with the concept of voltage, which is also called *potential*. For example, any electrical generating device—the two-century-old electrostatic machine, a storage battery, radio's microphone, the TV camera—all generate this pressure, or tension, we call a voltage.

Like the water behind the dam, an electrical potential is usually useful only when provided with a channel through which to flow. And because current will also flow between two similar voltages, either negative or positive, provided they are unequal, we had best use the concept of *difference of potential*.

Suppose, for example, the potential between one point in a circuit and ground is 12 volts, while between another point and ground is 20 volts. Both points are *negative* with respect to ground. But if we connect a circuit *between* them, with an ammeter in it, the meter will register a flow of current. For the 12 volts negative is really positive with respect to the 20 volts negative. There is a *difference of potential* between them.

When the power source supplies more than a single device, we have two basic choices in the way we connect them to that source—the "single file" *series* connection, or the "across the line" parallel connection.

All of our lamps and household appliances use the parallel connection, which means that each receives the 120 volts supplied by the power company. Being subject to the same voltage, each device will use an amount of current determined by its resistance ($I = E/R$). The power consumed is equal to the volts times the amperes ($P = EI$).

A toaster, for example, is a high-current user, and so it has a low resistance. One with a 1200-watt rating uses 10 amperes. ($P = EI$; $I = P/E$; $I = 1200/120$; $I = 10$ amperes.) Therefore, its resistance is 12 ohms ($R = E/I$). In contrast, a 40-watt lamp uses only one-third of an ampere. ($I = 40/120$; $I = 1/3$.) So the lamp must have 30 times the resistance of the toaster.

Now suppose we connect the low-resistance toaster and the high-resistance lamp together in *series*, and plug them in to the 120 volts. The current flow through the two devices will be governed by their *combined* resistance, which will be lower than it was for the toaster and lamp connected singly across the line. As there is no damming up of the current at any point in a series circuit, the current will be the same through both toaster and lamp. Because the toaster receives less than a third of an ampere, it won't function at all, and the lamp will burn with less than normal brilliance. Still, the high-resistance lamp will receive most of the power, because it is receiving the same current as the toaster, and must therefore have a higher voltage across it ($E = IR$). And $P = EI$. The rule is that in a series circuit, the voltage divides in direct proportion to the resistance of its various parts. This is an important consideration in many electronics circuits.

The filaments of the tubes in most radio sets and many TV receivers are connected in series so that the 110 volts of the power supply is divided among them. The higher a filament's resistance, the greater the voltage across it and the more power it consumes. Therefore, the tube that must handle more power than the others has a filament with a relatively higher resistance. The filament of the *power* tube (or tubes) next to the loudspeaker may have a resistance almost as great as the other filaments combined.

4

TWIST THAT DIAL, TURN THAT KNOB

"I HAVE heard from the family that the silk used by Henry was really a series of ribbons of silk obtained by the sacrifice on the part of his wife of her white petticoat. An electromagnet made by Henry and used in his experiments is still in existence, being treasured in the museum at Princeton University, and one can still see the white silk ribbons used in its construction, so I believe the story I have heard is true." (From an article in *Science,* published in 1932).

Joseph Henry, of gentle disposition, was happily married, and it is unlikely that Mrs. Henry objected to the deterioration of her wardrobe in promoting scientific inquiry. Nevertheless, a short time later, we find him insulating the wire for his coils by wrapping them with linen thread from the general store. One of his early horseshoe magnets, upon which he wound nine coils of linen-covered wire, all coils connected in parallel, lifted 750 lbs. with the current from a single galvanic cell.

Our discussion of Ohm's law, together with series and parallel circuits, prepares us for an explanation of why Henry used the parallel connection of the coils with a single cell. The cell potential was something less than two volts. This low voltage requires a low-resistance circuit in order to push sufficient current through it ($I = E/R$). When you connect nine coils of equal resistance in parallel, the total resistance will be only 1/9th the resistance of each coil alone. (A number of conductors in parallel is like a single conductor of their combined size, and, of course, the larger a conductor, the less its resistance. We can look at it this way: the larger a conductor, the more vagrant electrons available for the road.)

46

With parallel connection of the coils, Henry obtained a high-current flow from his single cell. The strength of the magnetic field is determined by amperes of current, multiplied by the number of turns, technically known as *ampere-turns*. This is what we need for the loudspeaker of our radio or hi-fi set, too. The output tubes, the power tubes, are built to deliver as much current as possible to the turns of the speaker's voice coil. The larger the audio frequency current, the more intense the voice coil's magnetic field, the further it moves the diaphragm, and the greater the volume of sound.

Henry probably added more of his chemical cells in parallel. That wouldn't have changed the voltage, but it would have made his power supply last longer. However, connecting the cells in series, as the three two-volt cells of our six-volt storage battery are connected, caused the cells' individual voltages to add. The higher voltage of the series-connected cells called for a different connection of the coils on Henry's magnet; meant connecting the coils in series also to raise their resistance. (The longer a conductor, the greater its resistance). A high-voltage battery would have forced so much current through the low-resistance, parallel coils that much energy would have been wasted in heat.

Joseph Henry, one of our greatest scientists, accomplished all this without being privy to George Ohm's research into the behavior of electricity. Nevertheless, he had Ohm's basic concepts in mind, as did Ampère in his experiments. Henry called his parallel-connected electromagnet, with its large current flow from a low voltage source, a *quantity* magnet; he referred to his series-connected, high-voltage magnet as an *intensity* magnet.

Now we come to an explanation of that term so often on the tongue of the hi-fi fan, *impedance match*. Impedance match is a matter of transferring power from one circuit to another with maximum efficiency. Henry discovered that in order to get the most power from a battery, he had to make the resistance of the attached circuit, called the *load*, equal to the resistance of the battery.

Any generator, chemical, magnetic, or tube, has an *internal* resistance of its own which is in series with the load. In other words, the power consumed is divided between the generator and its load. And only when the power is *equally* divided does the load obtain the maximum power that the generator is capable of delivering to it. This is what is meant by *matching* the load to the generator.

Suppose we have a 100-volt generator with an internal resistance of one ohm. We connect it to a load whose resistance is also one ohm. The total resistance in this circuit will be two ohms. Therefore, the 100 volts will push a current of 50 amperes through the load and generator ($I = E/R$). Fifty amperes at 100 volts means 5,000 watts of power ($P = EI$).

As the current is the same in all parts of a series circuit, and the voltage across any part is directly proportional to resistance, the one ohm of the generator and the one ohm of the load will divide the power between them. The load will have 2500 watts.

Substitute any load of less than a single ohm, or more than a single ohm, and figure out the power in it. The answer will always be less than 2500 watts. With the alternating currents of electronics the resistance becomes *impedance,* though the rule still holds.

Joseph Henry's coils, and the principles that emerged from his experiments, were later appropriated for the telegraph, the telephone, and finally for radio, TV, and other electronic applications.

The discovery of *mutual* induction between two separate coils by Michael Faraday, described in Chapter 3, was made independently, at about the same time, by Henry. However, Henry was first by several years in discovering *self* induction, which gave us the tuning coil. Here is an excerpt from his account of the discovery, published in 1832:

". . . When a small battery is moderately excited by diluted acid, and its poles, which should be terminated by cups of mercury, are connected by a copper wire not more than a foot in length, no spark is perceived when the connection is formed or broken; but if a wire 30 or 40 feet long be used instead of the short wire, though no spark will be perceptible when the connection is made, yet when it is broken by drawing one end of the wire from its cup of mercury, a vivid spark is produced . . . The effect appears somewhat increased by coiling the wire into a helix . . . *I can account for this phenomenon only by supposing the long wire to become charged with electricity, which by its reaction on itself projects a spark when the connection is broken.*" (Italics mine).

Note that final line. A coil, as well as a capacitor, can be charged up with electricity. We think of the electricity as being stored in Faraday's lines of magnetic force. We saw, in the case of the spark coil, how the primary's lines of force, as they rise and fall, induce a voltage in the secondary coil. However, as the spark that appeared when Henry broke the circuit revealed, the primary coil itself is also affected by the rise and fall of the magnetic field. Ergo, when the current in it is changing, a single coil *induces a voltage in itself*.

We are dealing actually with two voltages rather than one. How do the two voltages act in relation to each other? The self-induced voltage always *opposes* both the rise and fall of the original, or impressed, voltage. This explains the terms *counter voltage*, or *back electromotive force*.

When Henry closed the circuit that connected the battery with the long wire, or the coil (helix), the induced counter voltage opposed the battery voltage. The effect of this was to slow down the rise of current through the circuit. When he removed one of the wires from the cup of mercury, opening the circuit, the counter voltage was again in opposition to the impressed voltage; this time it tried to maintain itself and keep the current flowing.

The counter voltage induced when a circuit is broken can be much greater than the impressed voltage. For the counter voltage is induced by the collapsing of the lines of force surrounding the wire or coil, and collapsing lines move very quickly. We shall explain this later.

Now let's bring our discussion up to date by substituting for Henry's direct current, which he made and broke to get his effects, a current that is *continuously changing*, one which we call an alternating or oscillating current. As it rises and falls, reversing direction once each cycle, its voltage always will be opposed by the counter voltage. This opposition reduces the size of the current flow.

We see that the coil's *self-inductance*, designated by the letter L, offers resistance to an impressed a-c voltage. But instead of calling it resistance, we call it *reactance*, designated by the letter X. Both resistance and reactance are expressed in ohms. You determine the total opposition to the impressed voltage by combining the two to obtain the *impedance*, which we designate by the letter Z. The method of combining the two needn't concern us yet, but if you're curious, it utilizes the formula devised by Pythagoras more than five centuries before Christ: the hypotenuse of a right triangle is equal to the square root of the sum of the squares of the other two sides. Here the impedance supplants the hypotenuse. $Z = \sqrt{R^2 + X_L{}^2}$. We use X_L to indicate *inductive* reactance because, as we shall soon see, there is also *capacitive* reactance (X_C).

The increased reactance obtained by forming the wire into

a coil is explained by the concentration of the lines of force, permitting them to react upon all of the turns with greater effect. Therefore, inductance increases with the number of turns and their proximity. It can be figured from coil measurements, size of wire, et cetera, using a rather complex formula.

Henry's fat spark when he broke the circuit is explained by the extremely high speed of the collapsing lines of force. Here is the tip-off to the fact that *frequency* also plays a part in reactance. The inductance of a coil is always the same, but its reactance to a changing current increases with frequency. This explains a coil's very common use as a *choke coil.*

In the crossover network of a high-fidelity speaker system you will find a coil on one side of the circuit, the side that goes to the *woofer* or low frequency speaker. (See Fig. 66; page 201.) This coil's inductance is low enough to offer small reactance to the low audio frequencies, such as those under 500 cycles, but it offers increasingly greater reactance to the higher audio frequencies, keeping most of them out of the woofer.

At this stage, we realize that both coil and capacitor store charges of electricity, the former in its magnetic field, the latter in its electrostatic (electric) field. When carrying a changing current, both coil and capacitor create a counter voltage, or back electromotive force, that is responsible for reactance (X).

But the capacitor's reactance (X_C) acts in an opposite manner to the coil's (X_L). The higher the frequency, the *less* the reactance of the capacitor, so we use it to choke off the lower frequencies. In the hi-fi crossover network, the capacitor is placed in one of the leads to the *tweeter*, the high-frequency speaker, and also to the mid-range speaker, if one is used. (See Fig. 66; page 201.) It passes these frequencies with relative ease, but offers considerable reactance to the frequencies under 500 cycles.

In Fig. 66, the coil is connected in *series* with the first speaker's voice coil, as is the capacitor with the second speaker's voice coil. With an "across the line" or parallel connection, the effects of coil and capacitor are reversed. The coil, in effect, "shorts out" the lower frequencies, but its greater reactance to the higher frequencies forces them to stay in the line; the capacitor "shorts out" the highs while permitting the lows to pass on.

Much better filtering in a crossover network, with what is called a steeper *cutoff,* can be obtained by the use of both coil and capacitor together in each circuit, one in series, the other in parallel, as in Fig. 66.

What we are talking about, in technical jargon, are *high pass filters* and *low pass filters*. A coil in parallel with a capacitor in series passes the highs best. A capacitor in parallel with a coil in series passes the lows best.

A low pass filter is found in most radio, TV, and hi-fi sets for smoothing out the 60-cycle a-c power supply following rectification to direct current. The object of this filter is to pass a frequency so low that it is down close to zero. Unless this ideal is approached, there will be a noticeable hum in the speaker. The conventional arrangement contains two large capacitors in parallel, with a large magnetic core coil in series. (See Fig. 28; page 107.)

We realize now that in an alternating or oscillating current circuit, inductive reactance (X_L) has an opposite effect on the frequency to capacitive reactance (X_C). This enables us to use a coil or a capacitor, or a combination of coils and capacitors, for a filter. A combination of coil and capacitor can also give us a *resonant circuit* for accepting a *single* frequency, and discouraging all frequencies above and below it.

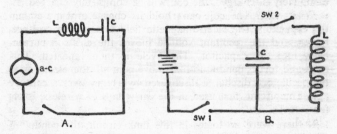

FIG. 8. (A) Series resonant circuit: the coil's reactance (X_L) and the capacitor's reactance (X_C) are equal at only one frequency, the resonant frequency. (B) The same is true for the parallel resonant circuit, called a tank circuit, the one most often used in electronics.

Fig. 8A shows a coil and capacitor connected together in series. We apply a low frequency, one that encounters considerable reactance in the capacitor, much less reactance in the coil. For this reason the net reactance to this low frequency is said to be capacitive. As we increase the frequency, the capacitor's reactance falls, the coil's reactance rises, until we reach the point where the two reactances are equal. They are in balance. The frequency that equalizes the reactances is the *resonant* frequency.

How do we explain this? Why does equalizing the two reactances produce resonance? Because when they are equal, their reactive effects in the circuit cancel each other out. How? Through their counter voltages.

The two counter voltages are directly opposite in *phase*. In other words, one is rising while the other is falling, and vice versa. Or, we can put it this way: when one is going positive, the other is going negative. So when they are equal, which is when the reactances are equal, they neutralize each other completely. At off-resonance frequencies, they only partially neutralize each other. Two voltages opposite in phase are said to be 180 degrees out of phase. This is illustrated in Fig. 10B; page 56.

Fig. 8B shows a parallel resonant circuit, the one most often used in electronics. A mechanical analogy best illustrates resonance in this circuit. First, we close Switch 1, charging up the capacitor with current from the battery. Next, we open this switch, then close Switch 2. This allows the capacitor to discharge into the coil. At the instant the capacitor is completely discharged, the coil will be completely charged.

What next? The coil can't hold its charge any more than the capacitor. The electromagnetic field around the coil collapses, and the resultant voltage moves the electron current back into the capacitor. The cycle is then repeated. This back-and-forth, pendulum-like movement of the electrons is the oscillatory discharge discovered by Henry over a century ago. The circuit, first used in the early days of wireless, is often called a *tank circuit*.

If there were no losses in this tank circuit, the pendulum movement of the electrons between electric field and magnetic field would continue forever. But the losses, through heat and some radiation, cause it to die out rather quickly.

The rate at which the electrons swing back and forth between capacitor and coil is the circuit's *natural,* or resonant *frequency*. Increase either the capacitance or the inductance, and the rate of the electron swing between them slows down. When you turn your radio dial, you vary the capacitance of an air capacitor until the tuner's resonant frequency matches the frequency of the desired station; in which case, the received voltage of the radio wave constantly replenishes the electrons oscillating between capacitor and coil. A TV receiver is tuned by switching in a complete coil together with the required capacitor, for resonance with the selected channel.

After tuning, the resonant frequency's voltage passes through a succession of radio tubes for amplification. The number of amplifying stages usually varies from three to eight.

To repeat, when capacitive reactance equals inductive reactance, their reactive effects cancel each other to provide resonance. Put more simply, we have resonance when $X_C = X_L$. How do we find the value of these reactances? Well, first we know that X_L increases with both the inductance (L) of the coil and frequency (f). To start with, let's say that $X_L = f \times L$, the f standing for frequency in cycles per second. But to make this formula work, for reasons we can't go into here, we must multiply by 2π. (π, the ratio of the circumference of a circle to its diameter, is approximately 3.142.) So the equation becomes $X_L = 2\pi fL$.

Since the capacitor's reactance acts contrary to the coil's reactance, the formula for capacitive reactance must be $X_C = 1/2\pi fC$.

The size of a coil (inductance), or of a capacitor (capacitance), is usually supplied by the manufacturer. The unit of inductance is the *henry,* after Joseph Henry; the unit of capacitance the *farad,* after Michael Faraday. These units are so large that electronics uses the *millihenry* (one thousandth of a henry) or the *microhenry* (one millionth of a henry), the *microfarad* (one millionth of a farad), or the *micromicrofarad* (one millionth of a millionth of a farad).

Abbreviations are always used; mh for the millihenry, μh for the microhenry, μfd for the microfarad, and μμfd for the micromicrofarad. When the time comes it will be interesting to learn from our galaxial neighbors on other planets exactly what units they use for inductance and capacitance.

Some resonant circuits are better than others. The quality is designated by the letter Q. It's mostly the resistance of the coil, appearing as heat, that lowers a circuit's Q. The Q is figured by dividing the coil's reactance by its resistance: X_L/R. If the dielectric (See Chapter 1) is air, the loss in the capacitor is usually negligible.

To keep its resistance low, the coil is wound with as large size wire as practicable. Because of the *skin effect,* wire size should not only increase with the amount of current flow, but with frequency as well. Skin effect is caused by the self-induced counter-electromotive force of the alternating current, which forces the current into the thin outer layer of the wire. This greatly reduces the wire's effective size and, consequently, its resistance.

Heat losses also occur in the insulation. For this reason, bare wire is often used, and the coil made self-sustaining to eliminate even an insulating core. However, this is not necessary with the frequencies used in broadcasting, at least in the receivers (medium wave).

Like a transformer, a resonant circuit can produce a *voltage increase*, if the losses are kept down sufficiently to yield a high Q. In a circuit with a Q of 100, for example, the voltage is increased 100 times. However, this voltage increase is usable only when directed to a tube's grid for amplification. The amplifier grid requires a voltage only; it uses a negligible amount of current.

The higher the Q, the *sharper* the circuit tunes, which means the more effectively it rejects (tunes out) unwanted frequencies. On the other hand, if a *band* of frequencies must be passed, as in the video amplifier of a TV set, some resistance may even be added to *broaden* the tuning; though this can be accomplished by the use of reactance with less loss of signal strength. Fig. 9 reveals why we use the terms *sharp* and *broad*.

With frequencies in the hundreds, as well as thousands of megacycles, so-called *lumped* inductance and capacitance, in the form of coils and capacitors, are seldom required. The inductance and capacitance in the connecting wires, or between the tube electrodes, may provide the tuning. We shall see an example of this in the dipole antenna of the Hertz laboratory transmitter of 1888, described in the chapter to follow.

The capacitor or coil in an a-c circuit has another effect, which we should touch upon briefly before taking up radio. The inductive reactance (X_L) of a coil causes the current to *lag* the voltage; the capacitive reactance of the capacitor causes the current to *lead* the voltage. Current and voltage *separated?* What kind of talk is this?

Current lag or lead simply means that the current and voltage rise and fall at slightly different times in relation to each other. This difference in time, difference in *phase*, explains the loss of power in a circuit with reactance . . . $P = EI$. And if voltage and current are somewhat out of step, neither can contribute its full value.

This out-of-phase business is something like a cow getting up. Her rear end rises first, followed by the front end. The

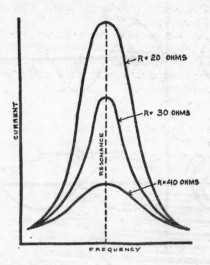

FIG. 9. Resonant peaks of a tuned circuit. The current is maximum at resonance, but falls off rapidly for frequencies either above or below resonance. The efficiency of the circuit, both as to sharpness of tuning and the amount of current flow, is governed by the amount of resistance in the circuit.

same with lying down. Horses don't go for this front end lag; they do it in reverse, with a rear end lag.

In a *purely* inductive circuit, the current would lag the voltage by a full 90 degrees, as indicated in Fig. 10A. However, a purely inductive circuit can't exist, because any coil or piece of wire always has some resistance, and the current and voltage in a resistance are always in phase. Therefore, the lag is always something less than 90 degrees. The same is true of current lead in a capacitor, though because of the reduced resistance, the phase difference can be closer to 90 degrees. A cow and a horse, getting up together, would be 180 degrees out of phase, just as the two counter voltages in a resonant circuit are opposite in phase when they cancel each other. (Fig. 10B.)

Perhaps the reader is wondering if the current lag or lead that results from the coils and capacitors in a cross-over network creates any ill effects. It's the *current* in the voice coil that causes it to vibrate; and what happens when the currents to the different speakers are out of phase with each

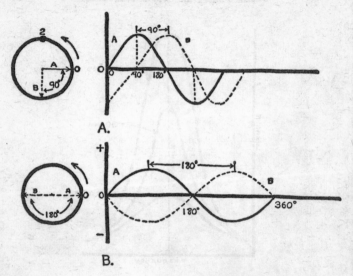

FIG. 10. The sine waves illustrate phase differences. (A) Assuming that the solid line is the voltage, and the dotted line the current, the current is shown *lagging* the voltage by 90 degrees. (B) Two voltages, or two currents, are shown 180 degrees out of phase. Since they are equal, they completely cancel.

other? The music from one speaker will either lag or lead the music from another speaker. This is one type of *distortion,* called *phase distortion.* Fortunately, the human ear doesn't seem to be bothered by it unless it's extreme.

I've often wondered whether certain animals, whose hearing is more acute than ours, are disturbed by phase distortion. My Great Pyrenees, Karlita, gets up and leaves the room often enough, when I play a Beethoven symphony, to make me suspicious. Of course, coming from the Pyrenees, maybe she just doesn't like German music.

This discussion of lag and lead should give us a firmer mental grip on the concept of resonance. Inductive reactance causes a current lag, capacitive reactance a current lead. At resonance, when $X_L = X_C$, the lag equals the lead, bringing the two currents together in phase. Not only do the counter voltages cancel in a resonant circuit, the impressed current and voltage rise and fall, reverse direction, then rise and fall again in complete harmony . . . like two cows of equal agility getting up and lying down together.

5

INTRODUCTION TO RADIO

THE Greeks lit fires on rows of hill-tops to signal the fall of Troy. Today we can talk all the way across an ocean merely by picking up a telephone receiver. These two methods of communication, bonfire and radio, have much in common: they both use the same waves; the only difference is in the *frequencies* of the waves.

Radio, from whose development all electronics largely evolved, sprang from the basic discovery, made late in the nineteenth century, that both light waves, and waves we can generate in an electric circuit, are *electromagnetic* in character. But before this connection between electricity and light could be established, each had to be investigated separately.

Is light instantaneous or does it "take time" to travel from point to point? What are the mechanics of seeing? Among the ancients, only the Greeks tried to answer these questions. Pythagoras and his followers believed that we see an object because it emits particles that proceed to the eye. Others, notably the Sophists, felt that it was more logical for the particles to originate in the eye. Empedocles, the eccentric philosopher from the temple city of Acragas, now Agrigenta, Sicily, who demonstrated the corporeality of air by means of a *clepsydra*—a form of water clock—sought a compromise. He maintained that a luminous body emits particles that are met by other particles from the eye. All of these views were influenced by the atomic theory of Leucippus and Democritus.

Empedocles guessed that light is not instantaneous but moves from point to point like the wind or a physical object. Aristotle approved, reasoning thusly: "Empedocles says

57

that the light from the Sun arrives first in the intervening space before it comes to the eye, or reaches Earth. This might plausibly seem to be the case. For whatever is moved (in space) is moved from one place to another; hence there must be a corresponding interval of time also in which it is moved from the one place to the other. But any given time is divisible into parts; so that we should assume a time when the sun's ray was not as yet seen, but was still travelling in the middle space."

During Roman times, the Dark Ages, and the Middle Ages, the nature of light stopped far short of the arena of burning questions. The subject didn't cross the threshold of pertinence again until the intellectual ferment of the Age of Enlightenment.

In 1676, the Danish astronomer Roemer confirmed Aristotle's reasoning. He also came close to figuring out exactly how fast the "sun's ray" travels in its long journey to our planet. The eclipses of the satellites of Jupiter take place on schedule, but Roemer noted that the schedule isn't quite as regular as had been thought. There is a difference, which, he assumed, depends upon whether the earth, in its journey around the sun, is between the sun and Jupiter, or whether the earth is on one side of the sun, Jupiter on the other. The maximum variation proved to be 21 minutes—the time required for light to traverse the diameter of the earth's orbit around the sun. Using Cassini's estimate of the size of this orbit, Roemer calculated that light must travel close to 192,000 miles per second. This is a mere 6,000 miles per second too fast, the present accepted figure being close to 186,282 m.p.s.

The great Galileo, who was brave enough to tackle almost any problem, had tried to measure the speed of light directly, using a lantern which he could open and close. Try to find the dimensions of an atom with your wife's tape measure, and you will realize how far off the beam Galileo was. Today's ingenious, highly complicated methods all come within a few thousand meters of 299,796,000 meters per second.

First to oppose the ancient particle theory of light with the wave theory was a contemporary of Newton, Robert Hooke. White light produces colored light when it shines through thin sheets, such as a soap bubble, an oil film, or a sheet of mica. This transformation is the same for all materials, and is determined only by the material's *thickness*. As the thickness lies between two limiting values, it is apparent that the color must be caused by some property of light that depends upon the distance traveled. This implies that light must travel in wave form.

Sound waves travel through the air, but in 1690 Christian Huygens wrote that light waves require some other medium. So he invented one: the *ether*. His new medium had to be composed of particles of perfect hardness and elasticity, like infinitesimally tiny billiard balls. Descartes also had postulated an ether, after the Greek ether, which was a primordial substance from which the universe is constructed.

Huygens claimed he had observed that light waves will bend around an object into the "shadow," like water waves around a piling. Unable to corroborate this experiment, called *diffraction,* Newton tried to account for all the properties of light through his "jet of particles" theory. It wasn't until early in the nineteenth century that the wave theory of light finally triumphed, as a result of Thomas Young's demonstration of *interference.* This experiment is now standard in high school physics classes. Two narrow beams of light are superimposed in such a way that some of the waves are 180 degrees out of phase, causing cancellation and a dark ring, while the in-phase waves are added to form a ring of light.

The wave theory of light ruled supreme for almost a century before it had to move over to make room for a new particle theory—*photons.* Light waves and radio waves, together with other electromagnetic waves, consist of particles and at the same time travel in wave form. This seeming contradiction didn't remain a roadblock in the path of modern physics for long. The physicist simply erected a DETOUR sign in front of it, and went on with his work.

Guided by faith in an ultimate simplicity, a plant carefully tended by the Greeks, the scientist always has striven, consciously or unconsciously, to unify the various manifestations of nature. Light, heat, electricity, magnetism, physical force —for the most part, seemingly unrelated—were finally assembled under one big tent with a banner across the front reading ENERGY.

The actors in this tent show can play any part. Physical force, for example, can quick-change into electricity; electricity into heat or light, or both, or into physical force; heat into electricity, et cetera. No energy is ever *lost* during these interchanges. For a time, Mr. Conservation of Energy doubled as Mr. Conservation of Matter.

Classical physics put this show on the road about the time that *Uncle Tom's Cabin* got its start; and it's still playing, though the MC's name has been changed to Mr. Conservation of Energy-Matter. Since Planck and Einstein, matter changes into energy, energy into matter. $E = mc^2$ tells the story— E for energy, m for mass, and c for the speed of light in centimeters per second.

When, in 1819, Oersted united electricity and magnetism, many natural philosophers speculated upon the possibility that light might be added to make a trio. We can find a beginning for this, as usual, among the Greeks. Empedocles attributed lightning to rays from the sun that had become imprisoned in the clouds.

Around 1744, J. H. Winkler, inventor of the amalgam rubbing pads for the eighteenth-century electrostatic generator, tried to deflect a beam of sunlight by means of an electrified glass tube. His contemporary, William Watson, who shared Franklin's theory of the electric charge, believed electricity must have some properties in common with magnetism and light. But it wasn't until a century later that any proof emerged.

In 1845, Michael Faraday, using a powerful electromagnet and a heavy borosilicate lead glass bar, succeeded where others had failed. He placed the glass bar across the lines of force of the electromagnet. Light, which had been *polarized,* was allowed to pass through the heavy glass, its rays parallel to the lines of force.

The electromagnet's lines of force twisted the polarized light waves out of their path very slightly. A reversal of the current flow through the coils of the electromagnet twisted the light rays in the opposite direction. This rotation of the plane of polarization by a magnetic field seemed proof enough of a direct relationship between magnetism and light.

The above experiment, together with Faraday's electric and magnetic fields, which travel through the ether as sound waves do through air, prompted Cambridge University's Clerk Maxwell to undertake calculations that led to the electromagnetic theory of light. Maxwell's now famous equations were followed by the Hertz experiments, and finally by the Marconi wireless.

Shortly before his death in March of 1857, Faraday wrote in a letter to Maxwell, "I hope this summer to make some experiments on the time of magnetic action, or rather on the time required for the assumption of the electrotonic stage round a wire carrying current, that may help the subject on. *The time must probably be as short as the time of light;* but the greatness of the subject, if affirmative, makes me not despair." (Italics mine.)

Unfortunately, Faraday was in very poor health, and he died before he could attempt the measurements of the speed of the electric and magnetic fields that travelled out from a current-carrying wire. It was left to Maxwell to accomplish this epochal experiment.

It is widely assumed that Maxwell, poised over his desk with pencil in hand, squeezed his famous equations through the mesh of his thinking by sheer force of genius. A genius he certainly was, but he had to start with some kind of measurement to support his theorizing, based on Faraday's suggestion in the above quotation. But instead of trying to measure directly the velocity of the magnetic fields, in the manner of Galileo's attempt to measure the velocity of light with his lantern, Maxwell first analyzed the problem mathematically. He came to the conclusion that the solution lay in the relationship between the force of the electric charge and the force of the magnetic current.

When charged, a capacitor possesses around its plates and terminals the force of an electrostatic field. Discharge the capacitor, and the force of the magnetic field appears around the circuit conductors. (The electrostatic field doesn't disappear, however; it co-exists with the magnetic field.) Now, the strength of the magnetic field is determined by the speed with which the current from the charge is moving. Maxwell figured that if the current were moving with the speed of light, the intensity of the two fields would be equal. In other words, his experiment consisted of balancing an electrostatic attraction against the magnetic repulsion of its current flow. He found that they were equal.

We can best understand the logic of this test from an experiment made subsequent to Maxwell's work by the American physicist Henry Augustus Rowland, at Johns Hopkins University. Rowland went all the way back to the Greeks' charged amber for his experiment. Tying a piece of the amber to a string, he whirled it around in the air, at the same time measuring the strength of the magnetic field created by the moving charge. He discovered that for the magnetic field's intensity to equal the intensity of an equal charge moving through a conductor, the amber must move with the speed of light.

Maxwell's famous equations received mixed notices. William Thomson (Lord Kelvin), whose mirror galvanometer had saved the Atlantic cable, approved of them. Most of the continental scientists, influenced by W. Weber's hypothesis, were skeptical. Weber's hypothesis was based upon Newton's assumption that direct forces act at a distance in a straight line, and that their electromagnetic forces propagate through space instantaneously, with infinite velocity, like the force of gravitation. The Faraday-Maxwell theory, as it was known on the continent, was more difficult to accept.

The titan of science in Germany during this period was Herman von Helmholtz, whose contributions to physics, chemistry, and physiology, particularly the physiology of sense perception, were enormous. It was von Helmholtz who first distinguished identical notes made by different musical instruments on the basis of their overtones or harmonics. The harmonics are multiples of the fundamental frequency: the second harmonic is twice the fundamental, the third harmonic three times the fundamental, et cetera. Von Helmholtz assigned Rudolph Heinrich Hertz, a former pupil who had become his assistant at the University of Berlin, to the task of checking out the Maxwellian equations.

It wasn't long before Hertz had translated Maxwell's page of complex equations into a circuit that *radiated* his electromagnetic waves. The "transmitter" had a spark coil, a spark gap and a dipole antenna. TV has made this same dipole antenna, with certain refinements, ubiquitous throughout most of the civilized world today. Hertz obtained his power from a battery of Volta's chemical cells.

As we know, the spark coil changes the battery's d-c to a-c, and at the same time multiplies the voltage hundreds of times. The primary and secondary cells of the spark coil in Fig. 11 are wound on a straight iron core, one coil on top of the other, with an interruptor at one end to continuously open and close the primary circuit from the battery.

When the interruptor's springy *armature* is in the position shown, the circuit is closed, permitting current to flow through the primary coil. But the instant this current flows, it magnetizes the iron core, pulling the armature away from its contact, which opens the circuit and stops the current flow. With the current off, the soft iron core ceases to be a magnet; this permits the armature to flop back, again closing the circuit, magnetizing the iron core, pulling the armature away from its contact, et cetera.

Let's say that the armature of our spark coil's electromechanical interruptor is capable of flopping back and forth 100 times per second. Connected across the secondary coil is a capacitor, a small tuning coil, and a spark gap between them. Now let's close the battery circuit with the key and watch what happens in this ancient contraption.

When the primary circuit is closed, a surge of high-voltage current from the secondary coil is received by the capacitor, connected directly across it. It can't go as far as the coil because of the open gap in between. The high voltage charges up the capacitor. But the distance across the spark gap is so calculated that when the capacitor's charge reaches its peak, the voltage breaks down (ionizes) the air in the gap, allowing the capacitor to discharge into the coil. This discharge is oscillatory, a swing back and forth between capacitor and coil, as in any tank circuit. The frequency of this oscillatory current is determined by the size of capacitor and tuning coil. Increase either the capacitance or inductance, and you lower the frequency.

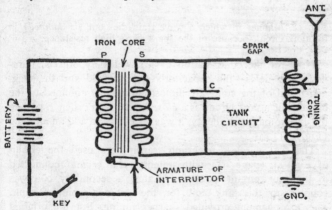

FIG. 11. Early spark coil wireless transmitter. The current from the battery, continuously interrupted by the vibrating armature, powers the spark coil. The induced alternating current (a-c), whose frequency is governed by the movement of the armature, charges the capacitor, C, with each half cycle of current. The capacitor discharges across the spark gap, setting up an alternating current, also called an *oscillatory* current, at the resonant frequency of the tank circuit. The resonant frequency can be varied by changing the number of turns of the tuning coil. Later transmitters used a *secondary* coil, coupled to the coil illustrated, to which the antenna and ground were connected.

There is one series of oscillations for each make and break (closing and opening of the circuit) of the interruptor, providing a tonal or audio frequency of 200 cycles per second.

Each series of oscillations, called a *wave train,* dies out quickly because of the high resistance of the spark gap (Fig. 12).

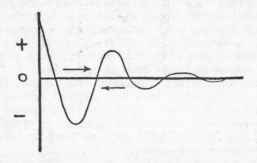

Fig. 12. Manner in which a wave train dies out (is *damped*) in the tank circuit because of the spark gap's high resistance.

Now for the confirmation of Maxwell's hypothesis. Faraday's two fields—the capacitor's electric field and the magnetic field of the coil—are almost completely confined to the circuit, for a tank circuit is a *closed* circuit. In order to radiate the electromagnetic energy, it was necessary to add an *antenna* (Fig. 13).

The first wireless telegraph transmitters used the circuit of Fig. 11 (page 63) with antenna and *ground* connected to the tank circuit as shown. Later, a secondary coil was coupled to the tank coil for the antenna-ground circuit. This arrangement provided sharper tuning. But the original Hertz transmitter was even simpler than the one shown in Fig. 11.

Hertz connected the spark coil secondary directly to the dipole antenna, with the spark gap in the center. The dipole itself, like any straight piece of wire, has some inductance; it also has some capacitance, mostly between its two wings. Sometimes Hertz attached metal balls or plates to the ends of his original dipole. Of course, without the coil and capacitor of the tank circuit, both inductance and capacitance are small, which means a very high frequency. Some of the Hertz frequencies were around 300 megacycles (millions of cycles). His apparatus may be seen today in the Science Museum at Munich.

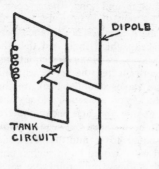

Fig. 13. This dipole antenna, basically the same as used today for TV, is connected to a tank circuit.

The two fields of the antenna circuit are not confined as they are in the closed circuit; they can spread out into space. And since the fields follow each other so rapidly, the magnetic field can't collapse on the circuit. We think of the succeeding fields as pushing each other out into space. Result: a radio or electromagnetic wave, first known as a Hertzian wave.

Fig. 14 reveals that the directions of the force of the two fields are at right angles to each other. They are also at right angles to their direction of travel. As the fields rise and fall, the electric field always reaches its peak intensity at the same instant that the magnetic field reaches its peak. Thus we have electricity and magnetism, considered separate and distinct phenomena for so many centuries, revealed as opposite sides of the same spinning coin.

The Hertz receiver was even simpler, consisting of a tuner and a "detector." The tuner was a single loop of wire, of the correct physical size for the right amount of inductance and capacitance required for resonance with the transmitter frequency. The "detector" was in reality a display device— a gap in the loop so small that it was scarcely noticeable. But if the loop were picking up energy, there would be a tiny spark across the gap. It developed that the spark was *so* tiny that Hertz could see it only in the dark. Did the received energy from the transmitter consist of waves, or had Maxwell been mistaken?

Hertz beamed the dipole transmitter at a piece of metal, a sheet of zinc six by 12 feet, mounted at the opposite end of the laboratory. When he moved his receiver back and forth between transmitter and zinc, there were points where the spark appeared across the gap in the loop, other points where there was no spark at all. The distance between these points was evenly spaced, which indicated that the radiated energy fanned out in waves, as Maxwell's equations had predicted.

The zinc sheet reflected the waves as a mirror reflects light. As a result, the waves doubled back on themselves, creating *loops* and *nodes*. (You see similar loops and nodes when you fasten a rope at one end and move the free end rapidly up and down.) If the distance between transmitter and reflector was correct—equal to a multiple of quarter wave lengths —the reflected wave would be in such a phase with the transmitted wave that they would reinforce each other. Otherwise a certain amount of cancellation would occur.*

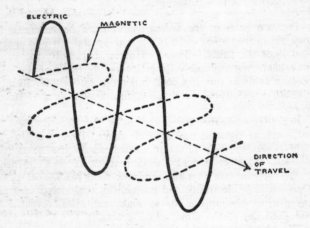

Fig. 14. Relationship between the directions of the two fields of the electromagnetic wave.

Were the new waves travelling at the speed of light? It wasn't necessary for Hertz to measure their velocity directly.

* Reflection sets up what is called a *standing wave*, or a *stationary wave*. Standing waves are a problem in transmission lines, such as the one between dipole antenna and TV receiver. If the impedance match to the dipole and the set is poor, the standing wave can cause *ghosts*. The ghost is a second picture, or even a third one, much fainter than the true picture and a little to its right. However, most TV ghosts are caused by another kind of reflection. A second wave from the transmitter, reflected from a mountain or steel building, reaches the receiver a few microseconds later than the direct wave, to form the second picture.

William Thomson, mentioned earlier in this chapter, had developed the formula for calculating the frequency of the oscillations from the amount of inductance and capacitance, both of which could be determined by physical dimensions. Knowing both frequency and wave length, the velocity is found by simple multiplication. Suppose, for example, the radiation comes from a circuit calculated to resonate at 150 megacycles. Measurement of the wave reveals it is two meters long. This means that the wave travels two meters each 1/150,000,000 of a second. During a full second, therefore, the wave would travel 150,000,000 $\times$ 2 or 300,000,000 meters, the velocity of light.

The Hertzian waves went sailing right through a wooden door. *Any insulating* material proved transparent to the radiation from his spark coil transmitter. Hertz fashioned a prism from cement that *bent* the waves, as light waves are bent by a glass prism (refraction).

To focus the new radiation, Hertz placed a parabolic reflector behind his dipole, the same type of reflector you see today on radar and microwave relay antennae. Radio astronomy uses monster versions of this device, such as the one at Joddrell Bank, England.

In order to discover whether the waves could be polarized like light waves, Hertz interposed a grid of wires between transmitter and receiver. When the grid wires were parallel with the dipole wire, the screen blocked the energy like a solid sheet of metal; when the grid wires were at right angles to the dipole, they had no effect upon the energy. A glance at U.S. rooftops today reveals that our TV transmitters must radiate a *horizontally-polarized* wave, or our horizontal antennas wouldn't intercept them.

Scientists as far away as India were soon repeating the Hertz experiments. Alexander Popoff, of Russia, built a receiver operated by static electricity to warn of the approach of thunderstorms. Edouard Branley, French physicist, and Britain's Sir Oliver Lodge, devised a much more sensitive "detector" of the waves, based upon a discovery by a scientist named Munk. It consisted of a small glass tube, filled with metal filings, which worked like a relay. The waves from the antenna caused the metal filings to cling together (cohere), thereby greatly reducing their resistance. When connected to a local battery, the lowered resistance increased the battery current, which operated a signaling device, such as a telegraph sounder or headphone.

The new "detector" was called a *coherer*. But because the filings tended to remain in their "frozen," low-resistance state, a *de-coherer* had to be added. Operating like the spark coil's interruptor, the de-coherer would tap the glass tube with a small hammer, causing the filings to de-cohere between signals.

Few of the scientists who pursued knowledge for its own sake gave much thought to the practical possibilities of the new waves. Some exceptions were Sir William Crookes, inventor of the Crookes tube, Sir Oliver Lodge, and the American Nikola Tesla. Electromagnetic energy travels in a straight line, and if the earth were as round as the scientists believed, the energy wouldn't get very far before it rocketed off into space. The range of a telegraph without wires would be limited to approximately 30 miles, depending upon the height of the antennae at transmitter and receiver.

In Italy, a young man who had been reading about the new low-frequency "light" overlooked the "line of sight" limitation. Apparently he believed that even a short-distance telegraph without wires would be valuable, for he had no way of knowing that certain frequencies do follow the earth's curvature to some extent.

Guglielmo Marconi was only 20 in 1894, six years after the Karlsruhe breakthrough, when he plunged into the experiments that soon lined his pockets with English gold and made the name Marconi synonomous with radio. Guglielmo's mother was a Jameson, of the famous Scotch-Irish family of whiskey distillers; his father was a wealthy business man. He carried out his early experiments on the family's country estate, near Bologna. The spacious, hilltop home, now drawn into the suburbs of the city, still stands, although it is unoccupied at the present time. His tomb is at the foot of the hill in the area over which he broadcast his first electromagnetic waves, working his way up to a record distance of 1¾ miles.

Marconi's only change in the laboratory instruments of the scientists was to lengthen the Hertz dipole. He extended one end of the dipole horizontally for a considerable distance; he grounded the other end. In effect, the earth substituted for one of the wings of a very large dipole. In 1909, Marconi was awarded one half of the Nobel prize for physics for this antenna.

Marconi's big antenna had much more inductance and capacitance than the little Hertz laboratory dipole. A large part of its capacitance was between the long, horizontal wires and the earth. As a result, the Marconi antenna resonated at a much lower frequency.

After he had removed his embryonic radio kit to England (his own Italian government disclaimed any interest in his work), the antenna continued to grow until its resonant frequency was in the neighborhood of 500,000 cycles (600 meters). These low frequencies produce currents in the earth, enabling much of their energy to follow its curvature for distances up to several hundred miles, depending upon the amount of power used. These *ground waves* encounter much less resistance over the sea because of the superior conductivity of salt water.

Marconi's low frequencies remained in favor, both for ship-to-shore work and limited trans-oceanic communication, for almost 20 years before there was any serious investigation of the merits of the higher frequencies with which Hertz and his fellow scientists had experimented. Then Marconi and others began playing around with frequencies ranging from two to 80 megacycles. Although these frequencies were thought to have a very limited range, they could be beamed conveniently as Hertz had beamed them, between relay stations. But there was a surprise in store for these experimenters.

As early as 1902, Britain's physicist, Oliver Heaviside, and Harvard's Dr. A. E. Kennelly, working independently, suggested that there might be something out in space that returned the man-made electromagnetic waves to Earth. This something proved to be what we now call the *ionosphere*. It returns enough of the energy for practical use of all frequencies up to around 30 megacycles (100 meters). Time of day, season of the year, and sunspot cycles are all factors in determining range and reliability. Long-range communication, both amateur and commercial, relies principally upon the frequencies between two and 30 megacycles.

The ionosphere consists of several layers of extremely rarefied air that has been ionized by the sun's radiation, mostly ultraviolet. The bottom layer is called the D layer, the middle one the E layer, and the one far out the F layer. During daylight hours, the F layer divides, the lower part about 100 miles above the earth, and the upper part ranging from 160 to 250 miles. Long-distance radio communication relies chiefly upon this F layer. Its relative proximity to the earth in the daytime accounts for the much shorter distance attainable in daylight hours.

When the layer of ionized gas is struck at an oblique angle by a radio wave, the wave is refracted (bent) to a degree that returns some of its energy to the earth. The energy in the frequencies above 30 megacycles are not refracted by the ionosphere to any very appreciable extent. These frequencies tend to pass straight on through the ionized particles.

There are exceptions. At certain times of the year, signals in the *very high frequency* band (30 to 300 megacycles) may be returned with considerable strength. A typical example is reported by the New York *Times:*

"On the evening of Nov. 8, 1957, the communications clerk in the Pima County Sheriff's office, Tucson, Arizona, called one of his patrol cars with an urgent message. No answer. But an operator in the Suffolk County, New York, Sheriff's office, 2300 miles away, picked up the message, relaying it on his own transmitter. This time the Tucson patrol car got it, rushed to the scene of the crime, and made an arrest. As the police car frequency is 39.18 megacycles, it should only be good for 'line of sight' transmission. However, there are periods during the spring and fall, lasting about six weeks, when the night-time ionosphere returns some of the very-high-frequency energy, making possible the so-called 'freak' reception. Calls are often picked up from police departments as far away as South America."

Since the two-to-30 megacycle range has become as overcrowded as a Moscow apartment, the engineers have developed a new communications technique called *scatter propagation.* By using extremely high power with a directional antenna, and several high-gain (ultra-sensitive) receivers at the other end (also directional) refraction from the ionosphere's E layer has been accomplished for frequencies ranging from 30 to 100 megacycles. In this way, distances can be covered ranging from 500 to 2000 miles.

Another scatter technique, called *tropospheric* scatter, has proved useful for distances up to 500 miles. The frequencies range from 100 to 100,000 megacycles. The troposphere is that part of the earth's atmosphere that extends upward from the surface for a distance of approximately six miles. There are no ionized layers, but the molecules of gas are not evenly distributed and tend to collect in areas called *blobs.* Because of variations in temperature and pressure, these blobs present a difference in dielectric constant to the electromagnetic wave that *scatters* its energy. Some of the scatter is certain to be in the desired direction, forward and down to the earth again.

Scatter communication is used by our military in Europe, between islands of the Caribbean, and between stations of our new long distance radar DEW (Distant Early Warning) line across Alaska and northern Canada.

6

AUDION TO PENTODE

ABOUT all we have left of the old Marconi wireless system, whose SOS captured the world's imagination half a century ago, is the tank circuit of the tuner (Fig. 22; page 90). Even though the open gap was greatly improved by the *quenched gap* of Germany's Max Wein, it couldn't be used for radio *telephony* because the oscillations were broken up into wave trains; in other words, it had a built-in audio frequency of its own that confined its use to the dots-n-dashes of radio telegraphy.

Even though the receiver's coherer was soon supplanted by improved detectors, the best one being the crystal galena, silicon, or carborundum (1906), they were still insensitive. This problem was finally, and beautifully, solved by the little *Audion*, which both detected and amplified the dots-n-dashes. The filament of this little tube gladdened the heart of the operator out of all proportion to its faint glow. Three Audions (one detector, and two audio frequency amplifiers) would double, triple, or quadruple the ship's working range —unless, of course, the static was severe, as was often the case on the low ship-to-shore frequency of 600 meters (500 kc), which is still in use for SOS calls.

Yet the Audion's aid and comfort proved to be the kiss of death to the spark transmitter. When supplied with proper voltages and some *feed-back* for oscillation, the tube proved to be a vastly superior *transmitter* as well. Unlike the spark gap, the tube generates a *continuous* wave, abbreviated *cw*, and it's a simple matter to impress upon a continuous wave a whole range of lower frequencies; the *audios* from 30 to 15,000, for example, or the much wider range of *video* fre-

72

quencies from the scanning process in the TV camera. This is *modulation*.

It is true that continuous waves were generated by the electric arc and by the alternator before the Audion, but it was impractical to modulate them efficiently. The tube also prepared the way for highly efficient fidelity modulation. In fact, the tube made possible the whole art of electronics, and reigned supreme as amplifier and generator until the advent of the little transistor in 1948.

In the receiver an audio or video frequency must be separated from the continuous wave frequency, often called the *carrier*. This demodulation process is called *detection*, and we'll have more to say about it later. None of the early wireless detectors—magnetic, electrolytic, or crystal—worked very well, and during the first decade of the century, the pressure was on for the development of a better one.

In 1883, five years before the Hertz experiments, Thomas A. Edison, seeking a way to prevent his carbon filament lamp from turning black inside, had sealed a wire in the glass envelope, close to the filament but not touching it. He noted that when this extra wire was given a positive charge, a small current would flow out of the lamp away from the hot filament. With a negatively charged wire, no current would flow.

We realize now that the positive wire attracted the negative electrons shed by the incandescent filament, but the electron theory was still 15 years away. In 1906, Ambrose Fleming in England adapted this "Edison Effect," as it has been called since, to a detector for wireless, substituting a metal plate surrounding the filament for Edison's wire. It was more reliable than the crystal but not as sensitive, and never became popular.

From this vantage point in time, one is tempted to account for the genesis of the Audion by paraphrasing the immortal Tinker-to-Evans-to-Chance double play as Edison-to-Fleming-to-de Forest. The British, loyal to Fleming, maintain that all de Forest did was add the grid. This is like saying that all the inventor of the chariot did was to add a pair of wheels to a sled.

After a false start with a modified coherer called a *Responder*, de Forest's search for a better wireless detector began in 1901 in his Chicago hotel room with a Welsbach gas burner. Operating a spark coil transmitter in the same room with the Welsbach burner, he noted that the sparks crashing across the gap caused one section of its mantle to burn brighter. He was elated to discover a connection between the spark set's electromagnetic waves and either heated gases or incandescent particles.

De Forest later realized that it was the *sound* waves from the crashing sparks that affected the gas mantle, but, fortunately, the discovery came too late to deter him from further experiments in this direction. In his New York laboratory, two years later, he tried a detector with two platinum electrodes extending into the gas flame of a Bunsen burner. This device had a glass tube instead of the Welsbach mantle. It picked up some signals from two ships in New York harbor, a feat that convinced him he was on the right track with his "incandescent gases."

Since the detector was primarily meant for use on shipboard, and ships are notoriously lax in piping gas to the radio shack, de Forest switched to electricity for his incandescence and used a carbon-arc lamp. The carbon lamp was too "noisy," and, for the first time, he tried an incandescent *filament*.

De Forest had read of the 1882 experiments of the Germans, Elster and Geitel, in which *ionization* had been obtained in a partially-exhausted glass tube by means of a heated filament. Unfortunately, he apparently was not aware of Thomson's discovery of the *electron*.

A filament must be heated in a vacuum and de Forest bought an old Sprengle mercury air pump. But before he could obtain a satisfactory vacuum the pressure of duties with his new wireless company caused him to abandon once more the search for a better detector.

In 1905, on the advice of his able assistant, Clifford D. Babcock, de Forest went to a lamp manufacturer for an incandescent lamp. He mounted a metal plate next to the filament. Although this combination of filament and plate was basically the same as in the Fleming valve, de Forest was on another tack. In addition to the filament-heating battery, he connected another battery between the plate and filament. He was hoping that with the electric field provided by this plate battery the received signal would act upon the ionized particles to change the *resistance*. Thus the signal would conceivably provide some measure of amplification through a relay action, as in the old coherer. The Fleming valve, on the other hand, operated as a simple rectifier; it passed only the positive halves of the alternating current cycle.

Even though this device worked poorly, de Forest obtained a patent on it, as Edison had patented the "Edison

Effect" without any conception of its usefulness. Finally, he hit upon the idea of more positive control of ionization in the tube: he wrapped tin foil around the glass, connecting the foil to the antenna-ground circuit. This gave him a third electrode, a *control* electrode, separate from the plate current through the tube, and for the first time he figured that he was "amplifying with gas."

The next step was to move the control electrode inside the tube, where it took the form of a plate, mounted on the opposite side of the hot filament from the original plate. Common sense soon directed him to place the new plate *between* the filament and the original plate. He had Babcock punch some holes in it so that it wouldn't block the passage of the "gas" between filament and plate. At that point, he was "amplifying with electrons," though he didn't realize it. A more practical control element proved to be a piece of wire, bent back and forth in the form of a *grid;* it soon replaced the plate with the holes (Fig. 15). This grid Audion was patented in 1908.

De Forest placed his bent wire grid as close to the incandescent filament as possible—a logical step. As we see it today, the closer the grid approaches the electron emitter, the *cathode,* the greater the effect of its surrounding electric field, put there by the incoming signal, upon the electron stream between cathode and plate. In other words, its control of the number of electrons reaching the plate is more positive. Suppose we probe further into this.

The electrons forced out of the cathode by the heat tend to "cluster" around it, like bees around a juicy clover blossom. The electrons alone don't travel far from the cathode, and they tend to obstruct those that follow. The net result is the creation of what is called a negative *space charge* around the cathode. The positive plate voltage can "syphon off" most of these electrons, but never all of them. A positive grid, on the positive half of the signal cycle, reduces the space charge, thereby adding electrons to those going to the positive plate. On the signal cycle's negative half, the negative electric field around the grid subtracts electrons from the plate current by reinforcing the space charge.

Fig. 16 shows the basic construction of a modern radio tube. The test of the "goodness" of a tube as a voltage amplifier is revealed by the relative effectiveness of the two voltages on the space charge, and thus on plate current flow through the tube. Because of its nearness to the cathode, the grid voltage always has the advantage. Let us say, for

FIG. 15. An early model of de Forest's Audion. This one had both twin grids and twin plates. The filament (cathode) was heated through the base. One of the top wires is connected to the grids, the other one to the plates.

example, that raising the plate voltage one volt increases the plate current by one milliampere (one thousandth of an ampere, abbreviated ma); whereas the grid requires only $\frac{1}{50}$ of a volt to increase the plate current by the same amount. This tube would have an *amplification factor* of 50.

Archimedes said that with a long enough lever, he could move the earth. De Forest's little bent wire grid comes as close to Archimedes' lever as anything we have devised so far on this planet. To take a signal example, it is now guiding giant missiles 8,000 miles to within a mile or two of a selected target.

The symbol for the amplification factor is the Greek letter μ (mu). (This Greek letter is also the symbol for *micro,* a prefix meaning "one millionth of.") Of course, in an actual amplifier, the amount of voltage amplification is always something less than the μ of the tube would indicate. The amplifier functions as a generator; and since any generator has an internal resistance, as we explained in Chapter 4, there is always some loss of power in it that lowers the amplification factor.

The above preliminary discussion of the tube reveals that the basic secret of electronics is extremely delicate *control*

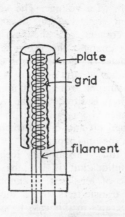

FIG. 16. Part of the plate removed from a modern radio tube reveals the filament (cathode) and the grid. When more grids are used, they are mounted in the same way, between the original grid and the plate.

—control of electricity by electricity, of a current flow by an electric field. Until the appearance of the little transistor, followed by other similar devices, this control was only attainable in a tube's vacuum, where the electrons have emerged from one solid conductor and are on their way to another. In other words, the tube freed the electricity from the solid "non-electric" of the early days—today's metal conductor—so that you could "get your hands on it," so to speak.

Control voltages on a grid, which are obtainable from any number of sources, natural or man-made, may have a potential of only a small fraction of a *microvolt*. The microphone converts sound waves into voltages of the same frequency. The *phototube* does the same for light. This little tube, widely used for self-opening doors, has a cathode coated with a photosensitive material, so that it emits electrons only when light shines upon it. The TV camera utilizes this material in converting the light and dark portions of a picture into voltages.

The phototube's cathode can be made sensitive to color changes. In testing a driver for intoxication, the police ask him to blow up a toy balloon. They discharge the contents of the balloon into a machine containing potassium dichromate, which alcohol turns blue. The phototube relates the degree of blueness, which reveals the amount of alcohol on the subject's breath, to a voltage. The voltage, when amplified and its current registered on an ammeter, reveals the suspect's alcoholic content, give or take a small beer.

Emotional stress changes the resistance of the skin. The polygraph, or lie detector, converts this change of resistance to a voltage, amplifies it, and records it on tape. Following amplification, the current from the changing voltage moves a coil in and out that is similar to the voice coil of a loudspeaker. But instead of being attached to a paper diaphragm, the coil moves a pen up and down, tracing the rise and fall of the current on a moving paper tape.

A succession of tubes can amplify feeble voltages millions of times. But amplification is not the tube's sole virtue. Through the magic of feedback, it can also generate waves for radio and TV transmitters and a thousand other operations. *Feedback* operates in this case to supply the oscillating tube with its *own voltage* (Fig. 17).

The air core transformer is the key to what takes place. Its primary coil is connected in the plate circuit, its secondary coil in the grid circuit. As Faraday discovered in 1831, any *change* in the volume of current, no matter how slight, in the plate circuit's primary coil will induce a voltage in the grid's secondary coil. This is feedback.

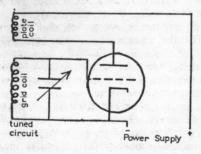

FIG. 17. Oscillator circuit, using inductive feedback between plate and grid coils. The frequency is regulated by the variable capacitor. The tube is a triode.

If the plate current rises ever so slightly, and the connections are such that the direction of the current through the coil is correct, the voltage induced on the grid will be positive. A positive grid voltage further increases the plate current, which, in turn, increases the grid voltage, which increases yet again the plate current, which—but you get the idea. It's like the cost of living index, except that in this instance, we are privileged to see exactly what causes the rise.

Of course, unlike the index, the plate current can't keep on rising forever. The supply of electrons emitted by the cathode is limited; and a *point of saturation* is soon reached when no amount of added grid voltage can cause any further increase in the plate current. Then the magnetic field around the coil collapses, and the plate current starts to fall. If a *rising* plate current induces a *positive* grid voltage, a *falling* plate current will induce a negative grid voltage, or else Faraday and Henry were mistaken a long, long time ago. The negative grid voltage encourages the plate current drop, which makes the grid even more negative, et cetera, et cetera, until the grid becomes negative enough to prevent *any* plate current from flowing. This is called the *cut-off point*. Then the current starts to rise again and the cycle is repeated.

The number of cycles per second—the *frequency* of an oscillator—is determined by the resonant frequency of the grid circuit's capacitor and coil. The feedback coil in the plate circuit is called the *tickler*. The shape of the wave, shown in Fig. 10 (page 56) is the same sine wave produced by the electromagnetic generator. Please note that it is not a half circle. The voltage rises very rapidly at first, then more slowly until it is scarcely rising at all as it approaches its peak. It falls in the same way.

This kind of feedback, if stopped short of causing the tube to oscillate, is called *regeneration*. Regeneration can multiply the sensitivity of a tube many times. In the circuit of Fig. 17, the amount of feedback can be controlled by varying the mutual coupling between the two coils. Regeneration is used in many communication receivers.

Feedback can take place through a capacitor as well as between two coils. A voltage on one plate of a capacitor induces a voltage on the opposite plate, as we explained in Chapter 1. In fact, a certain amount of this capacitive feedback is always present: it occurs through the tube itself. The plate and grid function as the two plates of a capacitor.

The early Audions were so prone to oscillate from this inter-electrode capacitance that it was difficult to find one that didn't. At five dollars per copy this was an expensive problem for the poor radio officer in the merchant ships, who had to buy his own Audion if he didn't want to suffer with the company-issued crystal set. We tried the cut-rate Japanese tubes, but the filaments didn't last.

Since 1929, we have used tubes with an extra grid, which

helps to reduce the inter-electrode capacitance with its feedback. Called a *screen grid*, because it screens the plate voltage from the control grid, it is connected to the plate side of the tube to keep most of the feedback away from the grid. In addition, this tube has a much higher amplification factor.

Placed between the signal grid and the plate, the screen grid carries a positive voltage, usually somewhat smaller than that of the plate. Nevertheless, since the screen grid is closer to the cathode, it is a great help in pulling electrons from the cathode's space charge. Although a few electrons are absorbed by the screen grid's mesh, most of them pass through it to the plate. Since the screen voltage takes most of the burden off the plate voltage in moving electrons through the tube, the control grid's "leverage" on the electron stream, *relative to the plate*, is much greater than before, thereby raising the amplification factor. The control grid's mesh also can be made finer, again adding to the amplification factor, which may be as much as 100 times that of the original three-element tube or *triode*.

In many circuits, the screen grid tube (or *tetrode*) has a serious defect caused by *secondary emission*. When accelerated electrons strike a conducting metal, their impact removes some of the metal's surface electrons. These are *secondary electrons*, and they are a useful phenomenon in the TV camera (Fig. 30; page 114). In an ordinary amplifier, the plate's secondary electrons may be attracted by the positively-charged screen grid, creating a second electron current through the tube—this one going in the wrong direction. As a result, a rise in grid voltage which should normally increase the plate current and, at the same time, the voltage across the plate circuit resistance, does exactly the opposite: it results in a voltage drop. Technically it's called *negative resistance*. Negative resistance can be useful, however, in causing a tube to oscillate without the need for feedback.

Secondary emission, which limited the usefulness of the tetrode as an amplifier, was remedied in 1930 by simply adding a third grid, placed between the screen grid and the plate. The third grid is connected to the cathode, which gives it a small negative charge with respect to the plate. The negative charge pushes back the secondary electrons to the plate before they can be picked up by the positive screen grid. Because it suppresses secondary emission, the third grid was named the *suppressor grid*. (See Fig. 18.)

The five-element tube—cathode, plate, control grid, screen grid, suppressor grid—called the *pentode*, is probably the most widely-used tube in electronics today. Most of the tubes in radio and TV sets are pentodes. The original tube, the triode, is still the standard transmitting tube, and is widely used for amplifying the audio frequencies, as in hi-fi installations. The symbol for the pentode is shown in Fig. 18.

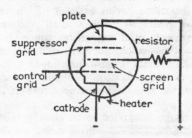

FIG. 18. The pentode, the symbol for the most universal of radio tubes. As it is common practice to operate the screen grid at a somewhat lower voltage than the plate, a *dropping* resistor is connected in series with it.

It wasn't until the period between 1912 and 1915 that the magic of the little Audion began to be appreciated. During this period, both de Forest and Edwin H. Armstrong were granted patents on a regenerative device that employed the Audion. Regeneration, of course, meant oscillation. The result was many long years of involved litigation between them, and between the companies that had acquired an interest in the device. The Supreme Court twice ruled for de Forest.

Armstrong later patented a frequency modulation system of broadcasting that made this method of modulation practicable. During his early experiments with the first Audions, he also devised a method of diagramming on paper the behavior of a particular tube when various voltages were applied to grid and plate. This diagram has become known as the *characteristic curve*. We could do worse than close this chapter with a short explanation of how to read the diagram. It should be a great help in understanding distortion in hi-fi sets, when we take up that subject in later chapters.

The characteristic curve is a map that shows at a glance how the plate current rises and falls when we change the voltage on the control grid. Fig. 19 is the characteristic curve of a triode for a given plate voltage.

The horizontal line represents grid volts, the vertical line shows the plate current. Note that with zero voltage on the grid, something like 16 milliamperes of current flow through this particular tube from cathode to plate. As the grid becomes more negative, the plate current drops, until a —16-volt grid shuts it off completely. This is the cut-off point, mentioned earlier.

The top of the curve begins to level off above 14 milliamperes and —8 grid volts. With a small positive grid voltage, the plate current would reach its maximum; a further increase in grid voltage would not change it very much. This is the point of saturation, also mentioned earlier. There simply aren't any more electrons available from the cathode —unless we heat it to a higher temperature at the risk of burning it out.

The signal voltage is applied to an already existing *steady* —8 volts on the grid. The —8 volts is called a *bias*. A bias is nearly always necessary in order to place the grid voltage on

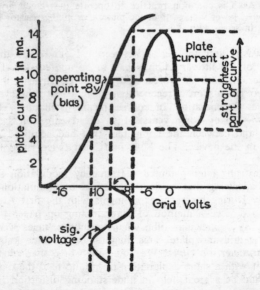

FIG. 19. We see a typical characteristic curve for a triode. The tube is biased at —8 volts, close to the center of the straightest portion of the curve, which guarantees minimum distortion. This is known as Class A amplification, which is almost universally required in amplifiers for audio frequencies. With Class B, amplifiers are biased below cutoff. Class B and C amplifiers are used mostly in transmitters.

the desired section of the characteristic curve. In this case—
and this is true for most amplifiers—the bias places the grid
voltage on the *straightest portion of the curve*. The signal
goes on top of the bias with no ill effects; the positive half
of the cycle makes the grid less negative, therefore relatively
more positive; the negative half makes the grid more negative.

The curve reveals that a two-volt positive half cycle of the
signal voltage, from —8 to —6 volts, causes a rise in plate
current from approximately 10 to 14 milliamperes; whereas
the two-volt negative half cycle causes a plate current fall
from approximately 10 to 6 milliamperes. The rise and fall of
plate current doesn't follow the grid voltage action with complete fidelity. It would only do so if the straightest part of
the curve were completely straight, something no self respect-
ing *curve* would be. The relationship between grid voltage
and plate current isn't *linear,* to use the technician's term.
Nonlinearity is at the root of almost all objectionable distortions in hi-fi amplifiers. We shall explore all this in detail in
the chapters on hi-fi.

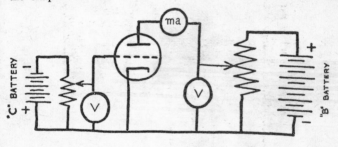

Fig. 20. Circuit used for plotting a characteristic curve. Both the
resistor across the C battery and the one across the B battery are
variable. Moving the arrow down reduces the voltage. First, the
desired plate voltage is selected, which registers on the voltmeter
V. Then a number of negative bias voltages are plotted against
the plate current, which registers on the milliammeter. A line
drawn between the points provides the characteristic curve.

Fig. 20 shows the circuit for plotting the characteristic
curve of a triode. First, the variable resistor across the "B"
battery is adjusted to yield the desired plate voltage. This
voltage registers on the V (for voltmeter). Then a number
of readings are taken for grid voltages by varying the resistor
across the "C" battery. These voltages register on the other

voltmeter. Each reading is plotted against the current flow registered by the milliammeter in series with the plate circuit. The "A" battery for heating the cathode is not shown.

You can use the same circuit to learn the tube's amplification factor. Simply compare the increase in grid voltage with the increase in plate voltage necessary to obtain the same increase in the plate current.

7

RADIO'S BEGINNINGS

ON APRIL 28, 1899, during a fog, the steamer *R. F. Mathews* rammed the East Goodwin Sands Lightship off the British coast. The lightship *wirelessed* the South Foreland Lighthouse, 12 miles distant, for help—the first distress call of its kind in history.

The operator had to spell it out, because CQD (Come Quick Distress) was still several years away, and SOS did not become a distress call until 1908. Jack Binn's historic call from the White Star liner *Republic,* when she was hit by the Italian ship *Florida,* off New York, January 23, 1909, was a CQD, as was the *Titanic's* call on the tragic night of April 14, 1912.

In 1899, Fire Island, off Long Island, and the Fire Island Lightship, 12 miles apart, were linked by U.S. Army Signal Corps's wireless apparatus. The warships of several nations welcomed aboard the new spark sets, and on September 21, Marconi arrived in the United States to report the America's Cup races off the New Jersey coast for the New York *Herald* by wireless. The signalmen with their flags won this early round hands down—or rather hands up and arms waving. Marconi's wireless didn't work.

During the following year, 1900, there was substantially greater progress in the new art. The *S.S. Kaiser Wilhelm der Gross* steamed out of Hamburg with wireless sputtering, the first sea-going passenger vessel to be so equipped. Headlines were made when her signals were heard *sixty miles* away. Nikola Tesla, the eccentric genius from Serbia who was the individual most responsible for our modern alternating current power system, wrote an article for *Century Magazine* in which he prophesied radar. He predicted that the new waves, "reflected from afar, may enable us to determine the relative position or course of a moving object, such as a vessel at sea." Hertz, of course, had amply demonstrated the reflectibility of the waves 12 years before. But the energy in the reflected wave from any appreciable distance is so small that radar had to wait for the amplifying tube.

The basic Marconi patent, granted in 1897, makes no mention of tuning, though Marconi was soon using a coil in the antenna circuit. On February 8, 1900, John Stone, American physicist and engineer, applied for a patent on a four-circuit tuner, two circuits for the transmitter and two for the receiver. Included was the closed circuit (the tank circuit), which we first discussed in Chapter 4.

Nine months later, the Marconi company applied for a U.S. patent on an equivalent circuit, British patent No. 7,777, which was granted. This famous 7,777 was responsible for the Marconi company's dominant position in U.S. wireless for the first two decades of the century. It wasn't until 1943, one month after Stone's death, that the United States Supreme Court acknowledged the priority of his patent. When millions are at stake, the goddess has been known to peek through her blindfold and tip the scales with her sword.

The event in 1900 that is most pertinent here is one that took place in December, at Cobb Island, Virginia, under the aegis of Professor Reginald A. Fessenden. Perhaps the most imaginative inventor in the field during those early years, Fessenden was a Canadian who came to this country to work for Thomas A. Edison. He later taught electrical engineering at Purdue, and the University of Pittsburgh, then at Western University of Pennsylvania in Allegheny City.

Fessenden was impatient with the dots-n-dashes of the spark transmitter with its built-in audio tone. He wanted a continuous wave that he could modulate with the voice. For his 1900 Cobb Island experiment he did use the spark coil, but he tried to raise the frequency of the electro-mechanical interruptor to a 10,000-cycle frequency, one to which the ear does not respond readily. While he couldn't make the vibrator work that fast, the voice modulation, heard a mile away, was, in Fessenden's words, "poor in quality but intelligible." It was also the first transmission of speech without wires.

In the same year, Fessenden began developing a high-frequency electromagnetic generator, an *alternator,* for his continuous waves. The problem was to build an ordinary generator whose armature could be spun fast enough for the necessary frequency and still stay in one piece. On Christmas Eve, 1906, one of his experimental machines, installed at Brant Rock, Massachusetts, carried music and speech to astonished ship operators at sea. This was probably the world's first radio broadcast. Six months later, Dr. Lee de Forest, in New York City, was transmitting music and speech from phonograph records with continuous waves generated by the electric arc of Valdemar Poulsen of Denmark.

Modulation was the road block for a radio telephone that used the alternator or arc. Bell's telephone had been saved by the early introduction of the carbon microphone, still in use. The resistance of a little cup of carbon granules, called the *button,* changes with the pressure of a metal diaphragm against it. The air waves of sound move the diaphragm in and out. There is no *generation* of current here, as in later electromagnetic and crystal microphones. A local battery circuit is simply varied by the carbon's changing resistance in step with the sound frequencies striking the diaphragm.

The old carbon mike has its drawbacks, including a steady hiss, caused by the current through the carbon granules, but it is rugged, and it produces a higher voltage than any of the mikes that have since been devised. The carbon mike still does a job with the feeble currents in telephone lines, but when it was subjected to the heavy currents in the antenna circuit of the early continuous-wave transmitter, the carbon granules tended to *pack,* even burn up. By connecting a number of microphones in parallel, in order to divide the current among them, and by encasing the microphone in a metal jacket, so that it could be cooled by circulating water, some improvement resulted.

Many other ingenious schemes were tried, both in this country and in Europe, but with only partial success. Then, in 1912–13, de Forest's "queer little bulb," the Audion, which had been a success as a wireless detector, suddenly began to amplify and oscillate.

Given a higher vacuum, enabling it to take a higher plate voltage and develop more power, and an oxide-coated cathode for improved electron emission, the Audion positively blossomed. Soon the alternators and the arcs, as well as the spark sets, were as obsolete, even for radio telegraphy, as the atl-atl. Modulation for radio telephony was no longer a problem. The feeble output from even an extremely sensitive microphone could be amplified until the power was sufficient for good low-distortion modulation of a powerful transmitter.

What is this process called modulation? A blob of clay in the practiced hands of the sculptor quickly assumes the appearance of a human head. We can think of the audio frequency as the sculptor, the radio frequency (continuous wave frequency) as the clay. The former molds the latter into its own intricate and subtle shape, merely by controlling the amount of power available for the oscillations.

Fig. 21 shows the radio frequency (r-f) current, followed by a simple sine wave audio frequency (a-f) current. The third section illustrates the former as molded by the latter. As the audio frequency rises and falls, it increases and decreases the amplitude of the radio frequency. Even a very complex combination of audio frequencies, such as those transmitted by a large orchestra and chorus, is effective in this molding process.

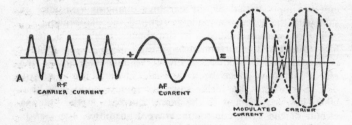

FIG. 21. Manner in which the radio frequency carrier current is modified by the audio frequency.

Note that the a-f appears on both the top and bottom of the r-f. In the receiver, one of these audio frequencies must be removed before the signal will operate the loudspeaker. This is the job of the detector.

Before exploring methods of modulation, let's pause a moment with the transmitter. The basic transmitter can be very simple—an oscillator, tube or transistor, coupled to an antenna.

Our first radiophone stations were built in 1915 by the telephone company whose engineers had just installed an improved version of de Forest's Audion in their long distance telephone lines. The stations were located at Montauk Point, Long Island, and Wilmington, Delaware, 300 miles apart. The equipment developed at these stations was installed in the U.S. Navy station at Arlington, Virginia, which, that same year, transmitted speech to Paris and Honolulu. The largest tubes available at the time were Western Electric's, rated at 25 watts. Five hundred of these little 25 watters were necessary to obtain 1½ kilowatts of modulated power in the antenna. The transmitter was modulated through the speech frequencies that varied the bias on the grids of the amplifiers. Grid bias modulation has not survived as a high fidelity method.

Those were exciting days in Arlington, when the country's first big radio telephone station, made possible by the invention of the triode, was established. The engineers must have blushed at the thought of having neglected this miraculous amplifier-oscillator-modulator for the preceding half dozen years. They made up for lost time. The grid modulation circuit was devised by a South African, H. J. Van der Bijl. Ralph V. L. Hartley invented an inductive feedback oscillator that has since borne his name. The designer of the transmitter, Raymond A. Heising, was soon to develop a superior method of modulation, in which the audio frequency was introduced into the power supply to an amplifier's plate circuit.

Fig. 22 explains this plate circuit modulation, which is still standard. The electron current in the microphone rises and falls in unison with the air waves of sound that strike its metal diaphragm. The mike is electromagnetic, built on the same principle as the dynamic speaker of Fig. 58 (page 182), except that it is operated as a generator instead of a motor. A small coil attached to the diaphragm vibrates in the fixed magnetic field of a permanent magnet, translating the motion into a fluctuating electron current. (A newer type, the *velocity mike,* vibrates the diaphragm itself, a corrugated metal "ribbon," in the magnetic field.)

From one to three stages of amplification are usually required to raise the feeble voltage from the microphone high enough to operate the modulator tube. Note that the modulator tube's plate circuit has in it the primary coil of a transformer, whose secondary is directly in the circuit that supplies power to the plate of the radio frequency amplifier. This *modulation transformer* couples the audio frequency into the radio frequency. The result of this coupling is shown in Fig. 21.

With the transmitter "on the air" and the mike dead, the secondary coil of the modulation transformer has a steady voltage across it from the steady current flow through it. Yet switch on the mike in a live studio, and watch what happens. Every positive half cycle of the audio frequency adds to the voltage across the coil, thereby increasing the current to the

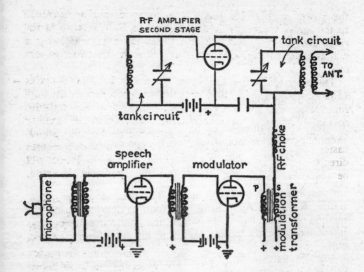

Fig. 22. Radio transmitter using amplitude modulation (AM). The power to the oscillating current in the second tank circuit is controlled by the audio frequency from the microphone. The audio frequency is coupled into the power circuit through the iron core transformer.

amplifier; every negative half cycle, by opposing the steady voltage, decreases this current. Thus the *amplitude* of the power that goes to the transmitter's amplifier is varied, in step with the audio frequency. As a result, the *amplitude* of the radio frequency oscillations also is varied. Today, this is referred to as AM, meaning amplitude modulation.

Because the modulator adds and subtracts current to and from the amplifier in equal amounts, plate modulation is also called constant current modulation. When I built my first broadcast transmitter, in 1922, the book in which I found the hook-up referred to it as "Heising modulation." The author of the book was Stuart Ballantine, a physicist, and one of our most brilliant inventors in the radio field, who died in 1944 at the age of 46. Among his more than thirty patents, his patent on *negative feedback* is of particular importance today in reducing the noise and distortion in high-fidelity amplifiers.

To return to Fig. 22, note the *radio frequency choke* (L), which prevents the radio frequency from feeding back into the power supply. It's really a one-way door; its reactance to the audio frequency is very low while its reactance to the radio frequency is much higher. This is an easy task for a coil because of the wide disparity between the two frequencies.

The lowest standard broadcast frequency is 550,000 cycles (550 *kilocycles*), and audio of 5,000 cycles is almost the limit for AM stations because their frequencies have been crowded so closely together to accommodate the maximum number of channels in the total space set aside for commercial broadcasters. The ratio is 110 to one. As a coil's reactance increases directly with frequency, the reactance in one direction will be 110 times greater than in the opposite direction. ($X_L = 2\mu fL$.)

As everyone knows, there is a newer type of broadcasting, called FM, for *frequency modulation*. Because there is less crowding of stations on the frequencies allotted to FM, its fidelity can be higher. With this type of modulation it's the *frequency* rather than the *amplitude* of the radio frequency that is varied over a limited range, naturally, to avoid interference with tuning. The FM channel is 150,000 cycles wide. Less static on the higher frequencies also makes FM more attractive.

Now let's jump to the AM receiver. First of all, the antenna signal must have considerable amplification if it is to operate a *transducer* larger than a pair of headphones. The second necessary function of the receiver is detection, which is another word for de-modulation of the signal. Finally, the above-mentioned transducer, in this case a loudspeaker, is needed to transform the electrical energy into acoustical energy. Any device that transforms one form of energy into another is a transducer.

The earliest mass-produced broadcast receiver was the TRF (tuned radio frequency) set, whose block diagram is shown in Fig. 23. This wonder-provoking machine included three stages of amplification of the r-f picked up by the antenna, each stage turned by a dial on the panel. There was also a rheostat (variable resistor) on the early models to control the voltage on the tubes' filaments. The r-f amplifiers were followed by the detector. There followed two stages of amplification of the demodulated signal, the a-f alone. Coupling between these audio stages was effected through iron core transformers. The final audio stage powered the speaker through another iron core transformer, called the output transformer, which is still in use in home audio circuits.

A major advantage of this early set was the absence of vocal commercials; a major weakness was the multiplicity of controls. Another problem was the feedback through the interelectrode capacitance that could cause the tubes to oscillate, thus chopping up the signal. The remedy was the addition of a circuit that fed back some of the plate's voltage to the grid circuit—out of phase with the voltage fed back through the tube. The cancellation prevented oscillation. This type of feedback is called negative feedback, or degeneration. It is universally used today in hi-fi amplifiers to counteract distortion and noise.

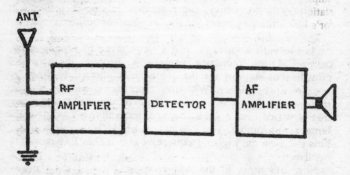

FIG. 23. Block diagram of the first tube radio receiver, the tuned radio frequency (TRF) set. Radio frequency amplification, with tuners between each tube, was followed by a detector, then an audio frequency amplifier of from one to three stages. Coupling between the audio stages was performed by iron core transformers.

The TRF receiver presented other problems too, most of which were solved in the mid-twenties by a brand new type of receiver, the *superheterodyne*. The superhet is pretty much today's standard receiver for all types of space communication, including both AM and FM radio, radar, and point-to-point radio. Its key is *conversion*. It converts the received frequencies to a single, lower frequency before much amplification, if any, takes place. The lower frequency, because it's *in between* the radio frequency of the transmitter and the audio frequency (or between the radio and video frequencies), is called the *intermediate frequency* (i-f).

Not the least of the superhet's virtues is simplicity of tuning. Because the i-f frequency is always the same, the i-f stages can be *fixed-tuned,* which means that no dial is required for adjustment.

The superhet story begins with radio pioneer Reginald Aubrey Fessenden, whose exploits we have already touched upon. Fessenden was primarily interested in voice-modulating his alternator's continuous waves (cw). However, dots-n-dashes carry much further than the human voice, especially on the static-ridden lower frequencies then in use. For the adoption of cw to radio telegraphy, Fessenden had to find a way to break it up into an audible frequency. His solution, which hasn't been improved upon since, was to *heterodyne* the cw with a second cw generated in the receiver. For example, if the two frequencies are 1,000 cycles apart, a clear, static-penetrating, 1,000-cycle note is obtained for the operator's key. This is the *difference* frequency.

The two frequencies both add and subtract—adding when the individual cycles are in phase, and subtracting when they are out of phase. Not only does a difference frequency of 1,000 cycles emerge, but other frequencies appear including one that is the *sum* of both frequencies. Actually, this is another example of modulation. Fessenden couldn't employ his discovery efficiently because he lacked a method for conveniently generating a local frequency, which today is readily obtained from an oscillating tube.

During World War I, German spark transmitters interfered with radio telegraph communication between the front line infantry and the barrage-laying artillery. A call went out to the Hammond Labs of John Hays Hammond, Jr., near Gloucester, Massachusetts, for a receiver that would be sufficiently selective to mitigate some of this interference. The Hammond engineers had been using intermediate frequencies in another connection, and they decided to apply it to this problem. Their solution was to raise the dot-n-dash audio frequency by heterodyning in the transmitter, raise it above the top frequency to which the human ear is sensitive. A filter circuit in the receiver could then be used to discriminate against enemy transmitters in favor of the new ultra-audio frequency. Finally, a local oscillator circuit heterodyned the ultra-audio back within range of the human ear. (I'm not sure this works.)

Performance checks on this system were made at the Signal Corps's Paris laboratory, early in 1918, under the direction of Captain (later Major) E. H. Armstrong. The available tubes amplified the low frequencies much better than the highs; at high frequencies, there was considerable loss in the inter-electrode capacitances, as occurred between the grid and cathode. Lowering the received frequency by heterodyning, before amplification, seemed a logical solution not only to Armstrong, but also to A. Meissner in Germany, L. Levy in France, and H. J. Round in Britain.

It occurred to Armstrong that this system might be applied to detect the high-frequency radiation from the spark plugs of German bombers in order to obtain a direction-finding circuit. Although the heterodyne circuit he developed in this connection was never perfected, because of the war's early end, the experience gained proved highly profitable.

On October 5, 1920, Major Armstrong sold his super-heterodyne and feedback patents to Westinghouse for $335,000. He was to receive an additional $200,000 if his feedback oscillator patent, which was being contested in the courts by de Forest, held up. It didn't. Armstrong's superhet patent claims in the U.S. were eventually awarded to L. Levy; his French claims, which discussed sensitivity alone, were upheld.

From 1913 until his death by suicide in 1953, Maj. Edwin Howard Armstrong was probably our most brilliant and productive radio inventor. During this 40-year period, he was involved almost continually in litigation over his patents, which earned him millions of dollars.

Soon after RCA, which shared cross-licensing agreements with Westinghouse, placed the first superhet receivers on the market, in the mid-twenties, a tube with an improved cathode was developed. The electrons did not derive directly from the heated filament; they were emitted by a nickel alloy sleeve, which surrounded the filament and was heated by it. Since then, the filament has been known as the *heater*.

This type of cathode explains why your radio or TV requires a little time to "warm up" before it starts to play. The filament first must heat the surrounding sleeve. This is not true, however, of the new transistor radio.

The sleeve-cathode holds its heat much longer than the filament—long enough, in fact, so that even if the current rises and falls 60 times per second, there is no critical falling off of electron emission. Today, the cathode can be heated by the 60-cycle a-c from a floor plug without causing a hum in the speaker. This was a big help in eliminating the old battery set. In the little a-c d-c set, the filaments are connected together in series, dividing the 120 volts a-c among them (pages 44–45).

8

SUPERHET

A READER of my *Electrons for Everyone* began a letter to me that opened with a little bouquet of praise, whose fragrance caused me to drop my guard. He then proceeded to lower the boom by confessing to one small reservation about the book. It's plain to see, he wrote, that any electronic circuit is made up almost entirely of capacitors, resistors, and coils—plus the tubes, if transistors haven't supplanted them. So why couldn't I explain, very simply, "just what each of these components *does* and be finished with it." What this man craves, obviously, is a book entitled *Electronics and Common Sense*—one I'm not just about to write.

How much more pleasant it *would* be if electronic circuits were no more puzzling than an auto engine, in which each component always has the same single task to perform. The carburetor mixes gasoline vapor with air; the piston compresses the mixture with an upward thrust; the spark ignites it, et cetera. But in an electronic circuit, the capacitor, resistor, and coil can do different things for us, determined by size, placement in the circuit (series or parallel, for example), and their placement in relation to each other.

Consider the little condenser, which we usually call a capacitor—that eighteenth-century scientific toy, in the form of jar, bottle, or phial, that so intrigued Ben Franklin. In today's electronics, it has more duties to perform than a carnival hand. If we could peek inside one, we would see that it always reacts to an applied voltage in exactly the same way. It stops cold a steady, unidirectional voltage (d-c), after taking an initial charge; and it offers opposition, called reactance (X_C), to a changing voltage in inverse ratio to the frequency. ($X_C = \frac{1}{2}\pi f C$).

Assuming our familiarity with the capacitor, and the fact that it outnumbers its nearest rival, the resistor, almost two to one in the superhet, let's use it as a guide to thread our way through the labyrinth of a superhet schematic.

The early radios demanded a roof-top antenna, usually a single wire stretched between poles. After receivers became more sensitive, and broadcast stations began to grow in power, a length of insulated wire, draped over a picture on the wall, would pick up enough voltage to operate most sets. Loops were installed inside the cabinet, and today, most small radios, and some big ones too, intercept the electromagnetic waves with a small coil wound on a magnetic core.

This core is something new, and comes from the same branch of science, solid state physics, that gave us the transistor. It is made from a non-conducting ceramic, in which oxides of iron have been mixed with one or more oxides of elements, such as barium and strontium. The material is molded (under heat at high pressure) into a rod called a *ferrite core*. A coil on one of these cores provides the most efficient small receiving antennae yet devised. The rod receives best when it is broadside to the transmitter. If you ever become lost in the woods some dark night with a transistor portable radio, you have a perfect direction finder.

In Fig. 24, we note that the capacitor C_1 is connected directly across the coil wound on a ferrite core. This gives us a parallel tuned circuit, the same as shown in the r-f amplifier

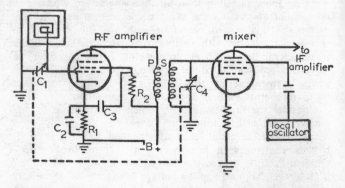

FIG. 24. Initial stage of radio frequency (R-F) amplifications, which is used with the better superhet sets before the incoming signal is mixed with the local oscillator frequency to obtain the intermediate frequency (I-F). The dotted line between the two variable capacitors indicates that they are tuned in unison.

of Fig. 22 (page 90). When the variable capacitor tunes this circuit to resonance with the selected station, the only opposition to the received wave's voltage will be the ordinary resistance in the conductors; because at resonance, the coil's and the capacitor's reactances neutralize each other. Therefore, the voltage across C_1 will be at its maximum, and this voltage is felt by the signal grid of the first tube, a pentode.

The current in the plate circuit's primary coil (P) follows the grid voltage as closely as the "straight portion of the curve" permits. (See Fig. 19; page 82.) It would be better if this coil were also tuned to the incoming frequency, but this would require an extra tuning capacitor, which would be awkward; you can obtain good results merely by tuning the secondary coil (S) of the transformer by C_4. The dotted line indicates that C_1 and C_4 are "ganged," which means they are on the same shaft so that they turn in unison. The maximum capacitance of each of these capacitors, when fully meshed, is around .003 μfd, depending upon the inductances of loop and secondary coil.

We come to the matter of *self bias*, which eliminates the need for a battery to supply this negative voltage. C_2 is the key to this stunt. When direct current flows through a resistor, one end of the resistor will always be positive, the opposite end negative. The electron current moves from the B— power supply, up through R_1 and the cathode, to the plate and back again to B+. This means that the bottom of R_1 will be negative, the top positive, as indicated.

The size of this voltage is determined by the size of the resistor and the amount of current through the circuit ($E = IR$). (In a series circuit, such as this one, the voltages divide in direct proportion to resistances.) A small resistor, say around 1,000 ohms, usually provides the correct amount of bias for the average tube. Note that the bottom of R_1 is grounded, together with B—.

As I have indicated, the lower end of R_1 is negative in relation to its top, and since the tube's cathode is merely an extension of this resistor, its lower end also will be negative in relation to the cathode. In order to make the *grid* negative with respect to the cathode, which is what we want, all we have to do is to connect it to the resistor's lower (negative) end, as illustrated.

Although the current through the tube from the B+ power supply is a direct current, a signal voltage on the grid causes a small amount of it to become a changing (rising and falling) current. This current rises when the grid goes positive on the signal's positive half cycle, falls when it goes negative on its negative half cycle. A current rise through the tube and R_1 would increase the negative bias voltage $(E = IR)$ tending to cancel out the signal were it not for C_2. C_2 also prevents the current fall from lowering the grid bias. Here is how it keeps the bias steady.

C_2 is called a *bypass capacitor*. It detours the rising and falling signal current around the resistor. To do this, its resistance (reactance) to the frequencies involved must be a lot lower than R_1's. The broadcast band is from 550,000 to 1,600,00 cycles (550 to 1600 kilocycles). A capacitor of .01 μfd has a reactance to the highest frequency in this band of approximately 11 ohms. A ratio of 11 to R_1's 1000 is more than satisfactory.

The reader may be puzzled as to how the capacitor can bypass the radio frequency around the resistor when their paths emerge at the same point. A more descriptive term for such a capacitor, one the British use, is *smoothing*—that is, the capacitor *smooths out* the frequency. A few cycles of the changing current charge up the capacitor; thereafter, it hasn't time to discharge (through the resistor) before subsequent cycles arrive. As a result, the capacitor holds the potential across the resistor constant, almost as constant as a battery voltage.

The final capacitor that concerns us here is C_3. We see that it is connected between the screen grid of our pentode and the cathode. It is essential to keep the bias voltage steady, and by the same token, it's important to steady the screen grid voltage. R_2 is a dropping resistor that provides the screen grid with a somewhat lower voltage than the plate. Most of the electrons pulled from the cathode by the voltage on the screen grid pass straight through the grid to the plate. Only a few are absorbed by the grid wires. However, these few electrons can cause trouble. They represent a frequency, determined by the rapidly-changing voltage on the signal grid, which can feed back through the screen into the power supply (B+) from where it can be passed on to the plates of the other tubes, causing all sorts of trouble, even oscillation.

C_3 is the solution. It offers a low-resistance path to ground for the energy in any *frequency* picked up by the screen grid. Together with R_2, it constitutes what is called a *decoupling* circuit or filter. (All amplifiers require decoupling, either separately or together.)

So far, we have investigated a single stage of tuned radio frequency (TRF), which is essential to a superhet if good

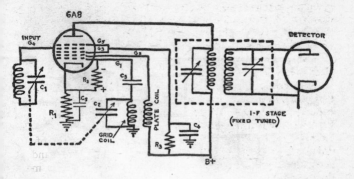

Fig. 25. Combined first detector and mixer tube of a little a-c d-c superhet, followed by a single stage of intermediate frequency (i-f) amplification.

selectivity and sensitivity are to be achieved. Next comes the problem of converting the tuned-in radio frequency of any station in the broadcast band to a single, intermediate frequency (i-f).

This, of course, calls for a local oscillator frequency to heterodyne (mix with) the tuned-in frequency. The i-f will be the difference between them. Originally, a separate oscillator plus a mixer tube was used, as indicated in Fig. 24 (page 96). However, in most of today's superhets, both oscillation and mixing take place in a single tube, which is often a *pentagrid converter*. This tube has five grids and needs every one of them.

Fig. 25 shows the front end of a typical little a-c d-c five-tube set with a pentagrid converter. Since it has no TRF stage (Fig. 24), this set needs only two variable capacitors on the shaft, one for tuning the antenna signal (C_1), and the other for tuning the local oscillator (C_3). It is followed by a single stage of intermediate frequency (i-f) amplification, as illustrated in the dotted square. After the detector, there are two stages of audio frequency (a-f) amplification (Fig. 26).

The incoming signal goes to grid No. 4 of this tube, which is a 6A8. Grid No. 2 functions as a plate for the oscillator circuit; instead of the usual spiral of wire, it consists of two rods, which enable it to absorb more electrons. Follow the B+ voltage through the plate coil to this grid-plate, and note that the plate coil is coupled to the grid coil, which is

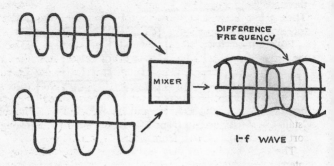

Fig. 26. Heterodyning (mixing) of two frequencies results in a difference frequency.

tuned by C_3. This is the same inductively-coupled oscillator illustrated in Fig. 17 (page 78), invented by Hartley.

The oscillator uses grid No. 1. This grid generates its own *bias,* and here is how it works. On the positive half of the oscillation cycle, the grid is made positive. A positive grid attracts some of the electrons to itself, the bulk of which pass on through the wires to the plate. These few electrons flow into C_3, charging it with the polarity indicated in the diagram. The electrons on C_3 have no place to go except back to the cathode through R_2. Because they leak off the capacitor through the resistor, the latter is called a *grid leak resistor.*

The time required for the accumulating charge to leak off is directly proportional to the product of the capacitance and resistance; the larger the product, the slower the leak and the higher the voltage on the grid.

The values for proper negative bias aren't too critical. A capacitance of .0001 µfd, and a grid leak of 22,000 ohms, will do the job, biasing down the tube below cutoff. Why so much bias? Wouldn't it be satisfactory if there were sufficient bias to work on the straight portion of the curve?

When two frequencies are "beat" together, *combined* as they are in the superhet, one of the frequencies that emerges is the difference frequency (Fig. 26). And note this point: the same variations appear on both the top and bottom of the waves. What we have, then, are *two* intermediate frequencies, which are quite obviously 180 degrees out of phase, one always rising while the other is falling, and vice versa. Unless we get rid of one of them, they will cancel each other out in the first stage of the i-f amplifier. We eliminate one by allowing only the positive halves of the i-f cycle to pass through the tube. This is called *rectification* or *detection*. Thus, the pentagrid converter must also operate as a detector, and it is often called the first detector.

With cut-off bias, which prevents the negative halves of the cycle from passing, another question arises: how does the signal grid (G_4) of Fig. 25 accomplish its work on the electron stream from the cathode with the result that it mixes with the local oscillator frequency? G_4 has a small negative bias from the bias resistor R_1 and its capacitor C_5. This bias "stalls" some of the electrons after they leave C_3, forming a sort of a second cathode, from which G_4 can draw.

The electron stream that reaches the plate, following the interruption by the signal frequency and oscillator frequency, contains not only these two frequencies, but also a sum and a difference frequency. However, the plate's tank circuit is resonant only to the difference frequency. This fixed-tuned primary of the first i-f stage is shown in the dotted line box of Fig. 25 (page 99).

The reader may be curious as to whether this combined frequency converter and detector also *amplifies*. Yes, it does, but the amount of amplification is only about one-fourth as great as it would be if the tube were operated as a straight amplifier.

Our little superhet has only a single stage of intermediate frequency amplification. Two fixed-tuned units are necessary because the i-f must have its entrance and its exit. Each unit is housed in a can, which you can readily distinguish by the two screws beneath the holes in the top. A similar can, without the holes, may contain the two large fixed capacitors for the power rectifier (see Fig. 28; page 107). The metal cover itself acts as a shield against stray fields, which could cause a hum or oscillation. The two screws in each can tune the primary and secondary circuits, either by varying a small capacitor, or by moving a "pulverized iron" rod in and out of the coil. The further movement of the rod into the coil increases the coil's inductance, and lowers the frequency. The service man may find this *realignment* necessary because of heat and aging. After the i-f, comes the second detector, whose explanation is now anti-climactic.

In the combined mixer-first detector, we separated the i-f from the r-f; now we have two a-f's, one on the top and the other on the bottom of the i-f, as indicated in Fig. 26 (page 100). Since the two a-f's are obviously 180 degrees out of phase, they will cancel each other out in the speaker's voice coil if we don't get rid of one of them. The most widely-used detector is the original tube, the diode. Current can only pass through a diode when the plate is positive; in this way, it blocks the negative half cycles. Selenium or silicon diodes are now replacing the more expensive tube.

The a-f is still a bit ragged with the i-f peaks. Therefore, we need another capacitor, C_1, for a bypass or smoothing job (see Fig. 27). C_1 is connected across the resistor in series with the detector diode. This resistor, R_1, passes on the a-f voltage to the first a-f amplifier through C_3. Since it is in series with the detector diode, and the secondary of the final i-f stage, R_1 must have a high value in order to capture a good share of the voltage (as we know, in a series circuit, the voltage is divided in proportion to resistance). 500,000 ohms (½ megohm) is common.

C_1 must smooth out an i-f frequency of 456,000 cycles, or thereabouts, without affecting the a-f. In other words, it must have a low reactance for the former, a higher reactance for the latter. A capacitor of .0025 µfd has a reactance to the i-f of 140 ohms. It's reactance to 5,000 cycles, which is about as high a frequency as the AM stations broadcast, and the little receivers are capable of receiving, would be roughly 90 times greater.

The variable resistor, R_1, also may be used as a *volume control,* as indicated by the arrow. The maximum volume is obtained when the voltage for the first a-f stage is obtained from across the entire resistor.

The diode detector circuit of our little set performs yet another function, automatic volume control (AVC). AVC means what it says—it automatically controls the volume so that when you tune from one station to another, it remains on a fairly even level. It doesn't raise a weak signal, but it levels off the peaks of loud signals.

Generally speaking, the greater the negative bias on a voltage amplifier (assuming that you stay above cutoff), the less effectively it amplifies. You'll notice that the curve of Fig. 19 (page 82) is not so *steep* when it is close to zero; this means that a grid voltage change has less effect upon the plate current. AVC provides a negative voltage for the grids of the r-f and i-f amplifiers that is governed by the signal strength; the stronger the signal, the greater this negative voltage.

AVC, then, is really automatic negative feedback. As signal strength increases, so does the current through the diode and R_1, which increases the voltage at its negative end. Here we find probably the first application of the kind of feedback

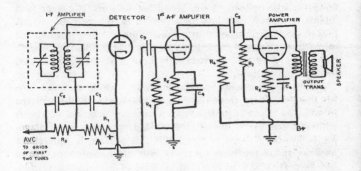

Fig. 27. Detector and audio frequency amplifier of a small a-c d-c receiver, showing the automatic volume control (AVC). In more modern sets, the first a-f amplifier is "housed" in the same tube as the detector, which is called a multipurpose tube. Both a-f amplifiers use a cathode bias. The audio frequency appears across R_1, the volume control, which is in series with the detector diode. The variable resistor, R_1, is called a *potentiometer*. The AVC uses the negative voltage from across R_1. C_1 smooths out the i-f; C_2 and R_2 smooth out the a-f. C_3 and C_5 are coupling capacitors between the tubes. The speaker is coupled to the power tube through an output transformer.

that has become the cornerstone of automation in industry. Any deviation from a desired pattern, any error, is made to produce a voltage. The voltage is fed back to tubes that correct the situation. When mechanical motion is involved, the amplifying tubes work through either solenoids or electric motors.

Of course, the AVC voltage, like the bias voltage, must be *steady;* it can't fluctuate with the audio frequency. Otherwise, it will tend to cancel out the signal because it is opposed in phase (180 degrees out of phase, as illustrated in Fig. 10B; page 56). So we need still another smoothing capacitor, C_2, across R_2. This capacitor must smooth out a much lower frequency than C_1, so it has to be larger. A value of .05 μfd is approximately correct with a large R_2. R_2 can be large (on the order of two megohms) because there is practically no current through it, and therefore little voltage drop across it. A large resistor can take a smaller capacitor, which is cheaper, the two comprising a decoupling circuit, previously described in connection with the screen grid.

After the diode detector, with its volume control and AVC,

we are dealing with the a-f alone, the same a-f we obtained from the microphone at the broadcast station. Of course, this audio frequency isn't an *exact* replica of the original. It has picked up a certain amount of distortion passing through the numerous tubes and circuits on its way to the speaker. And in Chapter 11, we will learn exactly how it is further distorted in the speaker itself.

The interelectrode capacitance in the triode is much less for the low audio frequencies, and the feedback through the tube is hardly a problem. (Capacitive reactance decreases with frequency). However, the pentode provides greater amplification than the triode and, for this reason, is universally used in radio receivers. Among hi-fi enthusiasts, the triode has its partisans, who prefer it because of its lower distortion. Distortion is the big problem in a-f amplifiers, though it can be reduced drastically in pentodes by means of negative feedback.

The audio amplifier grids must be biased in order to keep the signal on the linear portion of the curve. The bias is even more important than it would be for r-f and i-f amplifiers because the signal has had more amplification and the swing is wider. Cathode bias, described before, is commonly employed. Because its bypass-smoothing capacitor must offer very little reactance to such low frequencies, a large one is required. C_6 on the final amplifier—the power amplifier—may have a capacitance of 20 μfd, which is large enough so that its reactance to 50 cycles is only about 160 ohms. Cheaper sets may use a grid-leak bias on the first audio stage.

Now we come to the final task for our two-centuries-old capacitor. C_3 and C_5 are used as *couplers* between the tubes. Since they also are connected in series with resistors, the hook-up is usually referred to as resistance-capacitance coupling (R-C coupling).

If we look again at Fig. 27 (page 103), we see that C_5 also prevents the high-potential plate voltage from reaching the grid of the power amplifier. Thus, it also functions as a blocking capacitor. It readily blocks the *steady* d-c plate voltage, though it just as readily passes the *changing* signal voltages.

The higher the frequency, the less the reactance of any capacitor. Since we must pass a *range* of frequencies, these two capacitors introduce a problem. Our little superhet's range may be only from 100 to 5,000 cycles, but the range of the hi-fi amplifier is usually from 30 to 15,000 cycles. What we do is make the coupling capacitor large enough to pass the lowest audio frequencies without serious loss—its capacitance may vary anywhere from .001 to .1 μfd. We can't make it too large, or it might leak current to the grid and change the grid bias. At this point, I might add that though the higher audio frequencies pass more readily (with less loss) through the coupling capacitor, they encounter some loss in the tube itself through the grid-cathode capacitance.

Actually, the entire R-C network between audio amplifiers must be seen as a unit. This involves a review of series and parallel circuits, and we shall take up the problem again in the chapter on hi-fi amplifiers.

Let's take a look now at the final a-f amplifying tube. When we were considering the preceding amplifiers, we were concerned with the problem of transmitting as high a *voltage* as possible from one tube to the next, but the final amplifier's job is to deliver a maximum of *power*. It's not voltage alone, but voltage times amperes, that the speaker's voice coil demands (P = EI).

Thus the power tube's amplification factor is sacrificed in the interests of the voice coil's current flow. Its grid wire is more open-meshed, the electrodes are more rugged, and a higher voltage is impressed upon the plate. To develop greater power, and also to reduce distortion, the better radio receivers and phonographs use a *pair* of power tubes in what is known as *push-pull* combination, which will be discussed in Chapter 11.

The FM receiver also is a superhet. It operates in essentially the same way as the AM set I have been describing, except for a different type of detector and an extra circuit called a *limiter*. How does the limiter work? With FM, the a-f varies the *frequency* of the continuous wave; that is, the positive half cycles increase the frequency while the negative half cycles lower it. This variation takes place within certain definite limits. The amplitude of the continuous-wave or carrier frequency should not be changed. However, some amplitude modulation is bound to sneak in, and the job of the limiter is to remove it. The limiter consists essentially of a tube with a sharp cutoff, which is operated between the cutoff and the saturation point. As a result, all oscillations whose height is greater than they should be, because of amplitude modulation, are sliced off evenly, and as most of the noise is AM too, it also bites the dust.

AM's simple rectifying detector won't pass muster with FM. The FM detector must convert frequency variations of the carrier wave to amplitude variations. It must discriminate between a frequency *rise* and a frequency *fall*, and for this reason it is called a *discriminator*. It converts the frequency rise to a positive half cycle of current, the frequency fall to a negative half cycle. These positive and negative half cycles become the audio frequency, which goes through the usual treatment of two stages of a-f amplification ahead of the speaker, as in the AM set.

The discriminator circuit is a little too complex for detailed explanation. Basically, it makes use of the current lag with an inductive circuit, and a current lead with a capacitive circuit. The discriminator is tuned to the central or *resting* frequency. By making the tuned circuit inductive an increase in frequency from the positive a-f cycle creates a current lead that increases current through a diode; a decrease in frequency from the negative a-f cycle creates a current lag that increases current through a second diode. The audio frequency appears across two resistors in the diode circuits— the positive half across one, the negative half across the other. The output derives from both resistors in series. It's really a balancing act.

The superhet receiver, used in all branches of space communication and radar throughout the world, contains many of the basic circuits common to all electronics. We have noted the importance of the capacitor to these circuits. When connected across a coil, it gives us the tank circuit for tuning. Tuners are used in the antenna circuit, and as couplers between the r-f and the i-f amplifying tubes. Between the i-f tubes, they can be fixed-tuned.

Most tubes today are self-biased, thanks to the capacitor. Connected across a resistor in the tube's cathode circuit, the capacitor smooths out the changing current to provide a steady negative bias. The tube may be an r-f, i-f, or a-f amplifier. A capacitor used with a large-value grid resistor also can produce a grid bias; the combination of the capacitor and resistor (with its slow discharge) allows the capacitor to accumulate the required negative voltage.

When used as a decoupler, the capacitor operates in essentially the same fashion as a bypass-smoother, except that in this instance, its job is to bypass a frequency direct to ground, which prevents it from affecting the direct current supply circuit. If it were allowed to get into the power supply, it could be passed on to the plates of other tubes and create distortion.

When we ask the capacitor to *couple* two audio frequency amplifying tubes, as in the audio amplifier, it responds as best it can. Even though its reactance drops as the frequency

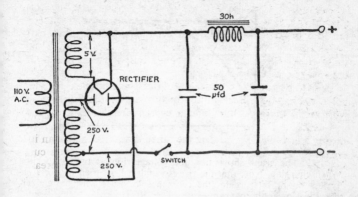

FIG. 28. Rectifier and filter circuit used with radio, TV, and hi-fi sets. The tube is a double diode. Each plate passes current through the cathode *only when it is positive*. Negative half cycles are blocked. Thus, when the top of a large secondary coil is going positive (and the bottom negative), current flows from the plate on the left, out through the cathode, into the set, and back through the center tap on the secondary. When the alternating current reverses, the *bottom* end of the large secondary coil goes positive (the top negative) and current flows into the set and back to the center tap from the plate on the right. The small secondary coil at the top supplies current to heat the cathode only. The low-pass filter takes most of the ripple out of the half cycles of direct current.

goes up, a good capacitor of the proper size will pass the audio spectrum (20-15,000 cycles) well enough. AM broadcasting limits the audio range to 5,000 cycles.

Another journeyman smoothing job for the capacitor takes place in the 60-cycle power supply circuit of radios and TVs. Here it must smooth out such a low frequency (60 cycles) that its capacitance must be very high so that its reactance will be low. Two capacitors of 50 µfd are commonly used. In addition, they need the help of a high-inductance iron core coil. As Fig. 28 shows, the coil is placed in series with the power supply, and the capacitors are connected in parallel. This low-pass filter is essentially the same as the one used in hi-fi's crossover network.

If we want to reduce the story of the capacitor to mathematical symbols, we need only two equations. $X_C = \frac{1}{2}\pi f C$ tells us that a given capacitor's reactance drops as the frequency rises; and for a given frequency, the larger the capacitor the less is the reactance. The second equation is $T = RC$

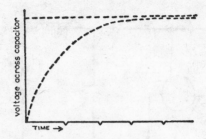

Fig. 29. The graph shows how the voltage rises across a capacitor when charging. Note that the rise is rapid at first, then progressively less rapid. The greater the resistance in series with the capacitor, the slower the charging rate (T = RC).

(the T is for time in seconds, R is for resistance in ohms, and C for capacitance in farads). The time in seconds is the charge (or discharge) time it takes a capacitor to reach 63 per cent of its full charge. (Theoretically, a capacitor is never completely charged.) Fig. 29 shows a typical curve for a capacitor charging through a resistance.

9

THE MAGIC SCREEN

Too many Americans seem to be convinced that electronics is strictly for the experts. It strikes me that a lot of us tend to shy away from even a simplified, basic explanation of the functioning of the electronic machine. We nonchalantly dismiss the subject with the remark that "all that stuff is too deep for me."

Many people apparently feel that their TV set can be traced back to Aladdin's lamp. How strange that television's operation should cast a spell over so many sophisticated citizens of one of the world's most technically proficient nations! This sense of wonder toward things technical would seem more appropriate to the New Guinea jungle or the African bush—until we recall that among primitive peoples, magic is often as much a staple as manioc, breadfruit, or sweet potatoes. The New Guinea head-hunter or the African bushman would most likely take TV in stride. Stick him in an elevator, or sweep him off the ground in a plane, and he would swell with consternation; but show him a picture hurled through space, and captured again on glass, and I doubt that he would bat an eye. The spirits he grew up with stage similar off-beat spectaculars at the will of the local shaman.

Now that loudspeakers enliven their huts with African hillbilly, the half million natives who live on the banks of the Congo in Northern Rhodesia call their radio set *kabulo ka kwabankain,* or "a small piece of iron that catches words in the air." It follows that when TV adds video to the audio, the theory will be stretched to include, "a small piece of glass that catches pictures in air."

Of course, us sophisticates ain't that naïve. We realize that it isn't actually the picture itself that comes hurtling through space. The magic for us lies in the transformation of the picture into "something else" before the launching, and the way in which the receiver reconverts this "something else" into a picture again. It's scant comfort to know that it's "all done with wires and electrons" when we are still mystified as to how the electrons are shepherded through the circuits.

Today's electrons most nearly approximate those ultimate, indivisible particles that Leucippus of Abdera imagined over 23 centuries ago. So far as we know, electrons *are* indivisible, and no particles are smaller, except some that must be in motion to possess mass, such as the photons of light and other electromagnetic radiation. Stop a photon, and it disappears, but we have a figure for the mass of an electron at rest.

As we explained in Chapter 2, ideally the electrons revolve around the nucleus of the atom, as the planets revolve around our sun. The comforting idea that small living things, including man, function in a manner analogous to the vast universe itself is as old as Anaximenes, of the Greek colony of Miletus (sixth century B.C.), if not older. Anaximenes compared the whole world with a single living organism. Wind is to the former, he wrote, as breath is to the latter.

Among later Greek philosophers who exercised this poetic parallel between the microcosm and macrocosm, Plato likened the soul of the universe to the soul of man. Aristotle developed the idea, along with his theory of circular motion, as a basic tenet of his philosophy. Even Harvey, discoverer of the circulation of the blood in 1628, likened the heart to the sun. And at one stage in the evolution of modern physics, the parallel seemed to have found beautiful confirmation in the solar system atom.

But this atom is no longer as simple as was first believed, especially since the physicist has been smashing its nucleus in his big machines. The resulting debris has yielded a growing list of smaller particles, some with a negative charge, some positive, some neutral. However, the first of the atom's particles to be discovered, the electron and proton, still provide a solid basis for the further development and the understanding of electronics. For example, in metals and their alloys, the atoms' outer orbit electrons often are held so loosely by their proton-charged nuclei that they are readily available for travel (page 30). Copper contains these wanderlust electrons in such vast quantities that in the magnetic generator it provides us with the bulk of our electric power.

Television harnesses these sub-tiny electrons to transform a picture into that "something else" we mentioned, then remake it into a picture again. The most important concepts we must salvage from the preceding chapters in order to understand this double metamorphosis are voltage and frequency.

Nearly everyone who has survived a brush with formal education has a fair working knowledge of these terms. Voltage is the electrons' electromotive force, their *potential* strength, which exists even when they are standing still as they are in a charged capacitor. Not until we give the voltage an outlet, a path, a *conductor,* do we get a useful movement of the electrons—a current flow.

The voltage doesn't cease with the current flow. In the simple series circuit the voltage divides itself among the elements in direct proportion to their resistance. As for frequency, it has nothing to do with the *velocity* of a current; frequency only tells us how fast the voltage rises to a peak, falls to zero, then reverses direction and rises and falls again. One of these complete routines, or alternations, is called a cycle. (See Fig. 10; page 56.)

The TV receiver harbors a broad range of frequencies, from the 60 cycles-per-second of the input power to the highest frequency of the video signal, which is something like three megacycles. The same applies to voltages. The signal from the dipole antenna may have a potential of only a few millivolts, or even a handful of microvolts. (In radar, fractions of a microvolt are not uncommon.) A steady, d-c positive potential of from 7,000 to 25,000 volts, depending upon screen size, is required to move the beam of negative electrons from gun to screen. The plates of the numerous amplifier and oscillator tubes take positive d-c voltages ranging from 70 to 300 volts. These plate voltages come from a transformer hooked to the power supply, which is followed by a rectifier and low-pass filter. (See Fig. 28; page 107.)

Compared with TV, radio is as simple to understand as a dinner gong. Speech and music are really sound frequencies in the air. If placed in the path of the sound waves, a metal diaphragm, attached to a coil floating in a magnetic field, vibrates in unison, the same as your ear drum does. The vibrating coil generates electrical frequencies (in itself) corresponding to the sound frequencies; and if we feed the electrical frequencies to a similar coil with diaphragm attached, the second diaphragm will vibrate in unison with the first to reproduce the sound. For the *generator* (microphone) is also a *motor* (speaker), and vice versa. Thus a couple of loudspeakers, their coils connected through wires, are all we need to operate a fairly efficient telephone.

Radio is faced with the additional problem of sending the sound frequencies, called the audio, *through space*. We have seen that this involves the generation of a higher, continuous wave frequency, called the radio frequency or carrier, which is *modulated* by the radio. In the same way, TV must transmit the audio frequency, though its biggest problem is light. Converting the lights and shadows of a *picture* to a frequency, and then using the frequency to recreate the picture, is somewhat more complex than is the case with sound. But it can be done, and in more ways than one.

Remember the old story about the tree that was so tall that it took a man and a boy all day to see to the top? What help was the boy? He kept the man's place during lunch and coffee breaks.

The story would have more logic and less humor if our eyes were made to view a scene one line at a time, the way a TV camera does. This contrived method of viewing is called *scanning*. A scene may be scanned in a number of different ways. By 1927, inventors in Britain and the United States were scanning the scene horizontally, using a mechanical system based upon Nipkow's scanning disc and the photocell, or the phototube. As you can see by a close-up look at your TV screen, today's electronic system also scans with horizontal lines. The "information" in these lines, the light and the dark, originate in the camera.

Like any good camera, the television camera boasts a system of high-quality lenses to bring the scene into focus, and at the same time let in as much light as possible. I don't have to tell you that here the similarity ends. Instead of film, the scene is focused upon what is known as a *photocathode*. Cathode is a term coined by Michael Faraday over a century ago. As I have pointed out, it means a negative electrode, *any* negative electrode, even the one on your car battery. But the *photo*cathode, as its name implies, is sensitive to light. It initiates the process that succeeds in converting the lights and shadows of the picture into electricity—with its two handles, voltage and frequency, though unlike a sine wave voltage (Fig. 10; page 56), the frequency is very uneven.

The type of TV camera in general use today for studio telecasting is built around RCA's *image orthicon* tube, which is not much larger than a pint beer bottle. A guided tour through this electron "bottle," the heart of TV, need not alarm you. It is no more difficult to understand, in a general way, than a full-rigged ship similarly encased in glass. Instead of masts, rudder, rigging, we watch for frequency, voltage, and the mutual reactions between positive and negative charges.

Our tour begins at the same end of the bottle that the light enters (Fig. 30). Note that the photocathode is directly behind the lenses, and that you can see through it. How does this small, thin, transparent plate generate electricity from the light focused upon it? Its secret rests in certain alkali metals, such as lithium, sodium, potassium, and caesium, which cling with feeble fingers to outer orbit electrons. You don't have to rub these materials with some other material to upset the electron balance. *There is enough energy in the photons of light to release electrons from their surface.*

The photocathode's inner face contains a thin coating of a caesium-silver compound. Since it is transparent, the light rays pass through it to release electrons from this coating. The number of electrons released from any point on the caesium-silver is in direct proportion to the intensity of the light reaching it. In other words, the brighter the light the more electrons *emitted*, to use the technical term. Because various areas of any scene reflect different amounts of light, the picture is reproduced in the relative densities of the emitted electrons.

The next step is to move these *photo*electrons down the length of the tube to another plate, called the *target*, without disturbing their relative positions in the stream. This suggests moving a smoke ring without changing its shape. A high positive voltage on a metal screen in front of the target attracts the electrons. They pass through the screen at high speed, and crash into the target, each electron still in its proper position in relation to the others.

The electrons' force of impact is sufficient for them to knock *secondary electrons* out of the target (page 80). This releasing of electrons by electrons is often a problem in conventional vacuum tubes but here we find it useful.

The target's loss of negative electrons leaves it positively charged. In other words, the lights and shadows in the scene before the camera have been "etched" upon the target in positive charges. The darker portions of the picture are represented by few, if any charges, the lighter portions by a high density of charges, with the grays in between. And as each high-speed electron has knocked out more than one secondary electron we have what amounts to amplification of the

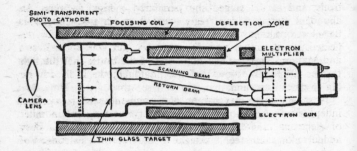

FIG. 30. The Image Orthicon, today's standard television camera tube. The image is focused upon the semi-transparent photo-cathode, which releases electrons in proportion to the brightness of the light that reaches it. These electrons, which have the picture in their distribution, are accelerated by a positive voltage, and crash into the thin glass target to release secondary electrons. The target's loss of secondary electrons leaves the image "engraved" upon it in positive charges. A low-velocity electron beam from the electron gun scans the target, line after line, from top to bottom, converting the image to a frequency. The voltage of this frequency is so low that it requires amplification by an electron multiplier in the neck of the tube. Conventional amplification of the frequency, called the video frequency, follows before it modulates the station's carrier frequency.

picture's electron image. This alone is enough to justify the use of the target.

The target, which is 7⁄8 by 1¼ inches, is made of a glass that is only one ten-thousandth of an inch thick. A pile of 40 targets would be equal to the thickness of a single sheet of paper of average weight. Because glass is an insulator, the positive charges stay put where they are "etched" by the loss of electrons; because of the target's extreme thinness, and the fact that no insulator is perfect, the charges move far enough to appear on the opposite surface, which faces an electron gun.

The electron gun has quite a long history. The Crookes tube, with which Roentgen took the first X-ray picture and in which J. J. Thomson discovered the electron shortly before the turn of the century, contained a cold cathode. To wrest electrons from a cold cathode, a little gas must be left in the tube. The negative cathode attracts positive ions from the gas, and their impact releases electrons.

A few years later, Ferdinand Braun used the first heated

cathode. A hot cathode greatly increased the supply of electrons, and at the same time permitted a high vacuum. An abundant supply of electrons made it much easier to focus the electrons in a narrow *beam*, which could readily be deflected for scanning, thus the electron gun was born. Braun and Marconi shared the Nobel Prize for physics in 1909, the former for his improved cathode ray tube, the latter for his big overhead antenna.

Next, Boris Rosing and A. A. Campbell-Swinton, working independently, devised a method of controlling the number of electrons in the beam during a scanning process. They added to or subtracted from the beam electrons by means of a changing voltage on a grid, placed in the path of the beam. Thus most of the basic research for the TV picture tube already had been accomplished before the inventors of all-electronic television came along; the major problem that remained was to develop a tube for the camera.

Roentgen and Braun were Germans; Rosing was a Russian; Crookes, Thomson, and Campbell-Swinton were British. We can work an American into the act in connection with the filament. Early filaments were made of nickel or tungsten. During the early years of radio tube development, U.S. physicist Irving Langmuir, another Nobel Prize winner (for chemistry) added thorium to the tungsten. A thoriated filament, and later a thoriated sleeve, proved a vastly-superior emitter of electrons.

To return to the image orthicon, the beam from the electron gun is focused by passing it through the magnetic field of a surrounding coil. The tip of the beam is brought to as sharp a focal point as possible on the target. A steady, positive voltage lobs the electrons up to the target. A low-velocity beam is required to avoid knocking secondary electrons from the target. And now we uncover the mystery concerning the method by which the beam from the electron gun transforms the scene before the camera into voltage and frequency.

The beam's tip scans the positively-charged target with 525 lines, each 1/30 of a second. (The way in which the beam is moved back and forth for scanning will be explained later when we cover the picture tube in the receiver.) Unlike charges attract, and the positive charges distributed over the face of the target remove negative electrons from the beam tip as it sweeps across, line following line from top to bottom. But the positive charges at any point on the target only remove enough electrons for their *neutralization*. The higher the density of positive charges, representing the lightest portions of the televised scene, the larger the number of electrons required for neutralization.

A negative voltage near the target returns the scanning electrons to the gun end of the tube. It is obvious that the number of electrons returned will vary; the electron voltage will fluctuate. To repeat, the fluctuations will be governed by the density of positive charges along each scanning line, which in turn is governed by the amount of light from the scene along that line. Thus, each line of voltage fluctuations contains a thin slice of the image. (See Fig. 37; page 124.) The full stack of 525 lines will hold the complete image. The voltage fluctuations are called the *video frequency* (video means vision).

Fresh from the target, the video voltage is very feeble. Before it can be used, a vast amount of amplification is required, first by an *electron amplifier,* operating upon the secondary electron principle (Fig. 31). Then conventional amplification will follow.

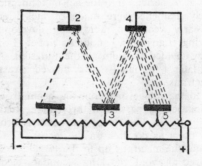

Fig. 31. Electron multiplier. The electrodes 2, 3, 4, and 5, called *dynodes,* have progressively higher voltages. The increasing acceleration of the electrons, as the voltages increase, causes them to release more secondary electrons when they strike the dynodes. The final electron current from dynode 5 is the sum of the initial current and the individual secondary electron emissions.

As the brighter, more positive areas of the image capture more of the beam's electrons, these portions create the valleys in the video frequency, and the dark portions form the peaks (Fig. 32). As a result, the picture is scanned upside down—it comes out a negative instead of a positive. This isn't serious because each time the video frequency passes through an amplifier, it is turned over. With an odd number of amplifiers, the picture emerges as a positive.

It is apparent that the video frequency increases with the amount of *detail* in the scene. For example, suppose we train our image orthicon camera on a country girl in a white dress standing in front of a red barn (red is dark on the screen). Each line of voltage will fluctuate enough so that you will note the difference between the white dress and the red barn. With few ups and downs in each line, the frequency will be relatively low.

Now, let's pack our rustic maiden off to the big city, fill her pretty head with a working knowledge of typing and shorthand, and find her a job. After a decent interval, let's dress her in a plaid skirt and polka dot blouse and set her on the knee of a boss who favors sharp sports coats. Move the camera in close, and you'll see what happens to the frequency. The voltage in each line keeps jumping up and down, like so many heart beats, in a vain attempt to capture all the detail.

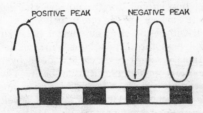

FIG. 32. A simplified illustration of how the light and dark areas of the image on the target are converted to a frequency. In the camera, the dark portions of the image create the positive voltage peaks, the light portions the negative ones. The frequency is applied to the TV picture tube as illustrated.

The amount of fine detail picked up by the video frequency in its fluctuations is determined partly by the "sharpness" of the beam tip. Let's suppose that the tip is sharp enough to note 400 variations in each scanning line of the scene depicted above. As there are 525 lines each 1/30 of a second, this would mean 6,300,000 variations per second (400 x 525 x 30). And as each variation produces a half cycle of alternating voltage, the frequency would be 3,150,000.

Another factor that governs the amount of detail noted by

the beam tip is the proximity of the lines. A lot of detail can be lost between the lines. Our 525-line television is barely adequate, though Great Britain manages to get by with 405 lines (wobbling the lines helps some), while France uses 809. Russia, Italy, Spain, Holland, Portugal, Germany, Greece, and Scandinavia all use 625 lines. Britain is now planning a changeover to 625 lines, which would not only improve the local "telly," but enable her to export her standard receivers to other countries.

Some of the detail in the high frequencies noted by the image orthicon's 525 lines always is lost on its way to the home screen. For example, a coaxial cable will chop off many of these higher video frequencies. The telephone company's network of relay stations is much superior in this respect. The best pictures are those televised live in the local studio. Given the same route to the transmitter, shows recorded on magnetic tape, a comparatively recent development, are better than films.

Among the formidable problems that had to be solved in the development of the image orthicon were the invention of a target thin enough for the positive charges to appear on the opposite surface, and a screen between the photocathode and target, with a sufficient number of holes so that the electrons could pass through without blurring the image. Most of the original camera research was done by America's pioneer television inventors, Russian-born Vladimir Kosma Zworykin and Philo T. Farnsworth, born and raised on a farm near Beaver City, Utah. Both men developed complete TV systems. Zworykin's camera tube is the *iconoscope,* still used for televising film; Farnsworth's is the *image dissector,* which has found a place in industrial television. Both cameras failed in the studio because they demanded lighting of too high an intensity, though the iconoscope is the superior in this respect.

If the video frequency reached your home over wires, your present receiver could be much simpler. This form of transmission is called closed circuit TV, a rapidly-growing service in schools and industry. Some day, sports events, feature movies, and new plays may be "piped" into your home for a fee. Pay-TV, which is being tested in several areas in America, *promises* to slay the commercial breathing dragon that dominates today's screens, and if true, it could well be worth the price. Whenever the blatancy, the syrupticity, or the chromium-plated charm of the sponsor's voice transcends my boiling point, I resolve never, under any circumstances, to purchase his product. Today I roll my own cigarettes, brew my own beer, bake my own bread, boil my own soap, tramp the woods for herbs, and import ripe, fragrant coffee beans direct from a mountain *finca* in Guatemala. Fortunately, I haven't yet been forced to deprive myself of bathroom fixtures, Kentucky blue grass seed, Erector sets, or Norwegian goat cheese.

Whenever the video frequency must be *telecast,* we encounter the same problems we met with radio broadcasting, plus a few extra ones. TV uses two carrier frequencies, one modulated by the video, the other by the audio. They are close enough together to be transmitted from the same antenna and received on a single antenna. The video is amplitude-modulated (AM); the audio is frequency-modulated (FM).

The TV receiver block diagram of Fig. 33 is revealing. Note that both carriers, audio and video, which together cover a wide *band* of frequencies, pass from the antenna into a superhet receiver with a stage of radio frequency amplification. The AM video and the FM audio stay together all the way to the first detector circuit. After the detector, the audio goes to its FM receiver, the video to a video amplifier, then to the picture tube. So far, we're on familiar ground: the set is a combined AM and FM receiver. However, I should point out that the signal also carries certain pulses, including *synchronization pulses,* inserted at the transmitter. These pulses are removed from the signal following video amplification, as indicated, and I shall describe their use later.

The picture comes to life on a glass screen: the large end of a funnel-shaped tube, called a cathode ray tube; or simply a CR tube. The name derives from the fact that the electricity passing through the early experimental models was first called a cathode ray, though in 1896, Sir J. J. Thomson demonstrated that the rays aren't really rays, but consist of particles of negative electricity, now known as electrons.

We come to the trick of moving the pencil-like electron beam rapidly back and forth across the screen. Scanning, both in camera tube and picture tube, is accomplished by *deflection.* The early picture tubes deflected the beam by means of rising and falling voltages on plates inside the tube (Fig. 34).

In his Crookes (CR) tube experiments, which led him to the discovery of the electron, J. J. Thomson used this electro-

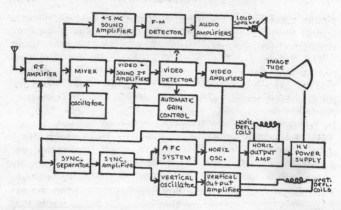

FIG. 33. A block diagram of a typical TV receiver.

static deflection. He later used the magnetism of coils to deflect the beam, though neither method was properly developed until after Braun's heated cathode had replaced the cold cathode, giving birth to the electron gun. Most modern

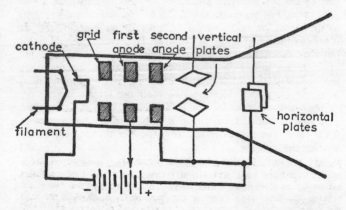

FIG. 34. Insides of a TV picture tube with the filament (heater), cathode, grid, and anodes for the positive voltages that move the electron beam through the tube to the screen. Also shown are the two sets of plates for electrostatic deflection. Voltages on the horizontal plates move the beam across the tube; voltages on the vertical plates tug on the beam so that each line is slightly lower than the previous one.

picture tubes deflect the beam by means of coils placed flat against the neck of the tube. Magnetic deflection is a more economical arrangement for the manufacturer.

The deflecting voltage (or current, if coils are used) in camera and picture tubes must rise and fall at the proper rate and the correct time. Fig. 35 illustrates a simple circuit for generating such a voltage or current. Its power source is a direct current, the kind that comes from a storage battery. Its key device is our old friend the capacitor.

The battery voltage moves current into the capacitor, C_1, *through the resistor*, R_2. As we know, it takes *time* to charge a capacitor, the amount of time depending upon the capacitor's size and the amount of resistance in series. ($T = RC$.) The rising voltage across the capacitor is connected to two of the plates in the picture tube that move the beam across the screen. A positive voltage on one plate pulls the negative electron beam at the same time that a negative voltage on the opposite plate pushes. This is *horizontal* deflection.

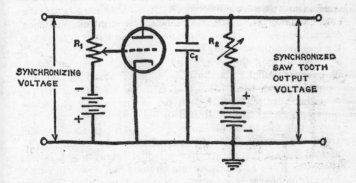

FIG. 35. Simple sawtooth generator for deflecting an electron beam in a TV picture tube. The beam is moved by the rising voltage across C_1 as it charges through R_2. At the instant the beam tip reaches the right-hand edge of the screen, C_1 discharges through the thyratron tube. A frequency of charge and discharge of 15,750 traces 525 lines on the screen each 1/30th of a second. R_1 regulates the bias voltage on the grid for the synchronizing pulse.

The capacitor and resistor must have the correct values so that the capacitor charges at the necessary rate, so many times per second. The capacitor must discharge at the instant the beam tip reaches the right-hand edge of the screen. As the beam must scan the screen 525 times each 1/30th of a second, the required frequency is 15,750.

The capacitor discharges through the tube, after which the beam is ready for another crossing. The tube shown is a *thyratron,* and is gas-filled. A gas-filled tube will not operate —that is, pass current—until the voltage across it has reached a certain value—high enough to break down the molecules of gas, causing *ionization,* and a quick rush of current. After discharge, the ionization disappears, and the tube again blocks the current flow, permitting the capacitor to charge again. The thyratron operates as a switch, controlled by a voltage.

However, it isn't possible to rely upon the voltage characteristic of the tube for the proper discharge time. The voltage of a *synchronizing pulse* is necessary. This pulse, generated at the transmitter, lands on the tube's grid, forcing it to discharge at the correct instant.

Note that the C_1 discharges directly into the tube and not, as on charge, through the resistor, R_2. As a result, the discharge time is only about 1/6th the duration of the charge time. The arrowhead through R_2 indicates that the resistor is variable. This is the Horizontal Size control. Reducing the resistance cuts down charging time, and thereby reduces the *length* of the scanning lines.

When we plot the charge and discharge voltage on paper, its shape resembles the business edge of a saw, which explains why a circuit of this kind is called a *sawtooth generator.* There are many different types of circuits for generating the sawtooth voltage illustrated in Fig. 36A.

We run into a non-linearity problem there, too. Fig. 29 (page 108) reveals that a capacitor's rate of charge can't be indicated by a straight line, like a ladder set against a house. However, if we add a correcting circuit (Fig. 36B), the early part of the charge is close enough to the linear form. The rest of the voltage rise across the capacitor is thrown away.

Non-linearity, as shown in Fig. 36A, means that the voltage rises quickly at first, and then begins to slow down. This indicates that the lines on the left-hand side of the screen move faster than those on the right-hand side. As a result, a circle will be pulled out on the left side, and flattened on the right side. Moving objects will grow fat or thin while passing through parts of the screen. On modern sets, the adjustment for correcting slight non-linearity is available only to the service technician.

The scanning lines on your TV screen, when the TV trans-

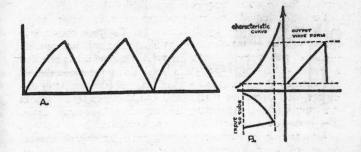

FIG. 36. (A) Shape of a sawtooth wave used for the deflection of the electron beam in picture and camera tubes. (B) Tube with the proper characteristic curve can "straighten out" the sawtooth wave so that it rises in a more linear fashion.

mitter is on but no picture is being telecast, are called *traces*. You can't count 525 of them, no matter how good your eyes are because some are lost in the machinery. If you are wondering why the beam tip doesn't make a *re-trace* on the screen when it returns to its starting point, the reason is quite simple: a big negative pulse, inserted in the signal at the transmitter, does the work. It takes place at the end of each trace, and blanks out the screen during the trace's return, which is called the *fly-back*. Since there are 15,750 lines per second, there must be an equal number of *blanking pulses* (Fig. 37A). The "blackout" is of such brief duration that the human eye can't detect it.

We haven't explained how a coil wound around the neck of the tube *focuses* the beam. The coil's lines of force are headed in the same direction as the electrons in the beam. (See Fig. 3B; page 35.) An electron sailing straight down through the middle of the coil toward the screen is fine; but all those that are out of line will start *cutting across* the magnetic lines, and they will begin to circle. However, as the electrons are being pulled along by the anode voltage (7,000 to 25,000 volts) at an extremely high velocity, they *spiral* rather than circle. By regulating the amount of current through the focusing coil, we can direct all of the spiraling electrons to a point on the screen.

Did your picture box ever present you with nothing but a single line of light across the middle of the screen? In a case

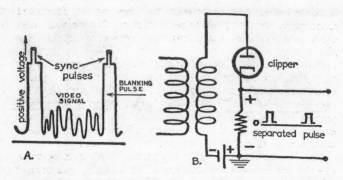

FIG. 37. (A) One of the 525 lines of the fluctuating video signal, which contains the light and dark information. At the end of each line is a highly-positive blanking pulse, which enables the trace to return to the other side of the screen without being seen. Mounted on top of the blanking pulse is a synchronizing (sync) pulse. (B) Diode clipper for removing the sync pulses from atop the blanking pulses. The plate of the diode is made negative enough by the bias voltage so that only the sync pulses are sufficiently positive to pass through it. Their voltages are taken off the resistor.

like this, you are receiving the complete picture, but it's in the horizontal traces only. You might as well try to read all the cards in the deck by looking at the back of the top card, a difficult feat even for Maverick.

In order to spread the traces out across the screen, we must have *vertical* deflection also. In the case of the single trace across the screen something is wrong in the vertical sawtooth circuit . . . usually a burnt out tube, in which case you can fix the set yourself if you can find out which tube it is. Unless tube filaments are connected in series, it will be the one that doesn't light up.

The voltage (or current) from the vertical sawtooth generator constantly tugs on the beam during the complete horizontal scanning, so that each succeeding line is a trifle lower on the screen than the preceding one (Fig. 38). The capacitor and resistor in this sawtooth circuit are chosen to provide a frequency of 60 cycles per second. Why not 30 cycles, if the horizontal scanning takes 1/30th of a second? The reason is *interlaced scanning.*

The scene before the camera is actually scanned with 262 1/2 lines each 1/60th of a second. And every alternate set of lines, or traces, goes *in between* the previous set. Interlaced scanning is necessary to prevent flicker. The vertical sawtooth generator also has a variable resistor to regulate the size control.

Now, if our introduction to the image orthicon, plus our quick tour of the cathode ray tube and its basic circuits, doesn't have your cortex crying for equanil, we'll look at *synchronization*, better known as "sync." Sync is the all important technique of keeping the beam in the camera tube exactly in step with the beam in the CR picture tube. Otherwise, strange things would happen to the picture.

Poor vertical sync can start the picture moving up or down like the old farmhouse roller towel. You usually can correct this fault by a control on the frequency of the vertical sawtooth generator available to the viewer. Many different evils can result from poor horizontal sync including "tearing" of the picture, a split picture, and black, diagonal bars across the screen. Controls for vertical and horizontal sync are those

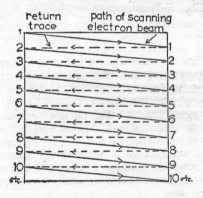

Fig. 38. The vertical deflection frequency of 60 cycles per second slowly moves the electron beam down so that each succeeding horizontal trace is a little lower than the preceding one.

marked Vertical Hold and Horizontal Hold. But local controls alone on a sawtooth generator would never work. The circuits must be policed by vertical and horizontal sync *pulses*, inserted in the signal at the transmitter.

The horizontal sync pulse is placed on top of the blanking pulse, which means that it comes at the end of each line.

(See Fig. 37A.) It is usually removed from the signal after one stage of video amplification, as indicated in Fig. 33 (page 120), then *clipped* from the blanking pulse.

Clipping is a very simple process that involves bias. You can use a diode, its plate negatively biased down below cutoff. (See Fig. 37B; page 124.) Then only the very top of the blanking pulse, the most positive part, which is the sync pulse, will be positive enough to pass through the tube. Thus, only a current flow that rises and falls like the little pill-box appears in the cathode circuit. When a triode or pentode is used for clipping, its *grid* is biased the correct amount below cutoff, and the pill-box appears in the plate circuit. Grid or cathode bias may be used.

After clipping, the horizontal sync pulse is amplified and fed to the horizontal sawtooth generator. In the case of the simple generator of Fig. 35 (page 121), the positive half cycle of the pulse would land on the grid. With the generator operating at a frequency even close to 15,750 cycles, this positive pulse, arriving at the end of each line, would *trigger* the tube, forcing the capacitor to discharge exactly on time.

Today's TV sets use a more complex arrangement to prevent interference from the sync circuits, and obtain even better stability. The gas-filled thyratron has been replaced by a vacuum tube, either triode or pentode, that *oscillates*. The oscillating frequency is kept directly on the beam by means of an automatic frequency control (AFC) circuit. The circuit is rather complex, but I'll give you the basic theory.

When the sync frequency varies ever so little, some retuning of the oscillator circuit must take place *quickly*. This corrective tuning could conceivably be done by varying a capacitor, the way we tune a radio, but this would pose an electro-mechanical problem—something more than formidable. So it's done with electrons alone.

In Chapter 4, we said that in a capacitive circuit, the current leads the voltage, while in an inductive circuit, it lags the voltage. So a little current lead (or lag), added to or subtracted from an oscillating circuit, will vary the frequency in the same way as changing the amount of capacitance would. Many AFC circuits use a discriminator, the kind found in FM receivers, whose output at the *correct* sync frequency cancels out; however, a higher or lower sync frequency upsets the balance and inserts a dose of current lead (or lag) in the oscillator. Current lead (or lag) may be obtained from a reactance tube—but we'd better let the details go or, so far as many of you are concerned, I'll be throwing nothing but junk.

The sync pulses that keep vertical deflection honest are somewhat different. The transmitter inserts a series of vertical pulses at the end of each "half picture," or 262 ½ lines. Since it is the same height as the horizontal pulses, the same clipper will remove them from the signal. After clipping, they must be separated.

The simple high-pass filter of Fig. 39A separates the horizontal pulses. At the same time, it changes the square pulse to one with a needle point—best for triggering the horizontal oscillator.

The vertical pulses, 60 per second as against the horizontal's 15,750, are much wider across the top. Through the simple filter of Fig. 39B, the wide pulses flow into the capacitor (which is larger than the one of Fig. 39A) through the resistor. After the required number of pulses, the voltage of the capacitor has built up to the point where the tube's plate voltage triggers the vertical oscillator. This low-pass filter is widely-used in many branches of electronics. It is called an *integrator* because it integrates (combines) the pulses fed to it.

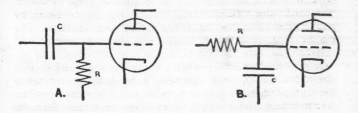

FIG. 39. (A) Illustrates the filter used with horizontal sync pulses, which trigger the horizontal sawtooth oscillator. (B) Integrating filter for the vertical pulses, which trigger the vertical sawtooth oscillator.

The vertical blanking pulse is so long that it blots out 20 lines of the picture. As it occurs once at the end of each "half picture," called a field, 40 lines are lost. So your picture actually has only 525 − 40 = 485 lines.

Our final moat to cross guards the mystery of the re-creation of the image on the glass screen. How do the 485 lines of the video frequency that rise and fall with the light and dark portions of the picture (see Fig. 32; page 117) come alive on the glass?

The coating on the inside surface of the glass screen is aluminum. But underneath this very thin aluminum coating is another coating of a substance known as a *phosphor*. A phosphor is luminescent. It emits cold light when struck by certain particles, such as X-rays, ultra-violet rays, or electrons. The aluminum backing merely serves as a reflector to increase efficiency. The electrons pass straight through it, causing the phosphor beneath to glow. The number of electrons in the beam governs the brightness of the glow along each trace.

The phosphor used in TV tubes is the kind that glows for no longer than 1/30th of a second after the beam tip passes. This permits a brand-new picture to be traced with each scanning—no overlaps. It is obvious that the fluorescent screen functions in a manner opposite to the camera tube's photocathode. The photocathode converts light to electricity; the picture tube's screen converts electricity to light. It is said that in 1904, Germany's A. R. B. Wehnelt first coated a CR tube screen with a phosphor.

Directly in front of the cathode of the electron gun is a cylinder with a small hole in its center. The electrons are "threaded" through this hole. The cylinder is the *control grid*. It acts like a door to let the beam's electrons through to the screen, a door that is opened and closed by means of a voltage (Fig. 40).

A negative-bias voltage on the control grid permits enough electrons to pass so that traces can be made of approximately the correct shade of gray—for background illumination. The operator can control background illumination by means of the Brightness control. The knob is connected to a resistor. Twist it one way, and the grid becomes more positive, brightening the screen; twist it the other way, and a more negative grid darkens the screen, even to the point of cutting off all the electrons in the beam, which leaves the screen totally dark. The adjustment of the Brightness control also must be correct so that the voltages of the blanking pulses can darken the screen at the end of each line and each field. Otherwise, the retraces will show.

Let's tune in a station and watch what happens. The very feeble video voltage from the antenna, whose trip through

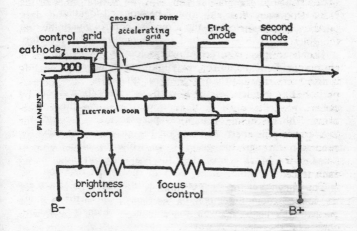

FIG. 40. Another view of the interior of the TV picture tube shows some of the controls. Each line of the video frequency, illustrated in Fig. 37 A, passes to the control grid, regulating the number of electrons in the beam along each trace. A positive rise increases the number of electrons, lighting the screen at that instant; a negative fall decreases the number of electrons, darkening the screen.

the superhet has amplified its strength several million times, goes to the control grid of the picture tube. As I have explained, this grid is negatively biased. The video voltage positive half cycle makes it less negative, or, in effect, positive; its negative half cycle makes it more negative. As the beam tip moves across the screen in sync with the beam along the target in the camera tube, the voltage of the video frequency increases and decreases the number of electrons in the beam, changing the degree of brightness along each trace. With perfect synchronization, the 485 lines of light and shadow stack up to at least a reasonable facsimile of the original scene.

The operator also can control the strength of the video signal, the amount of amplification. This is done through the control marked Contrast. The stronger the signal, the greater the difference between the peaks of positive and negative voltage, and the greater the contrast in the picture.

(See Fig. 37A; page 124.) Contrast and Brightness controls react upon each other and should be adjusted together.

Above a certain point, the Contrast control overloads the amplifier and wipes out detail in the picture. It is therefore advisable to keep it turned down as low as possible in favor of the Brightness control. Daylight watching, however, requires more contrast than is good for quality.

The 30 complete pictures each second (30 *frames*, 60 fields) give the illusion of motion, the same illusion the movies obtain with 24 pictures per second. TV's pictures aren't nearly as sharp as those on the movie screen, and the gradation through the grays from white to black isn't as satisfactory. Much is missed by the beam tip and much is left between the scanning lines, both in the image orthicon and picture tube. Yet the images on the little rectangular screens are still convincing enough for the sale of pills with as many as five "medically proven ingredients," which attack your illness in "five different ways."

This, of course, still falls far short of old Dr. McGintaw's famous remedy, sold at a loss by the "advertising agents" in front of a tent, which had as many as 365 "medically proven ingredients," consisting of roots, herbs, and berries, gathered from the headwaters of the River Nile, the only river in the world that flows north (move in closer folks, the dancing girls are next). But the image orthicon is still young.

And so we see that there really is no need to hold ourselves "superior" to the electronic intricacies of television. The TV set is simply a big box filled with tubes, coils, capacitors, and resistors, whose job is to tune, amplify, detect, oscillate, filter, and clip for two purposes: (1) to convert an FM audio frequency to sound in a loudspeaker; (2) to convert an AM video frequency to lights and shadows on the glass screen of a CR tube. Simple?

Okay then, it's just a big box with a piece of iron and a pane of glass for catching words and pictures in the air. Credit those native Rhodesians with a knack for going straight to the heart of the matter.

10

THE TRANSISTOR

WHEN over a century and a quarter ago, Georg Ohm laboriously worked out his theory for the behavior of electricity in a solid conductor (now called Ohm's law), he didn't overlook temperature. He found, for example, that the resistance of a metal increases with temperature. When you switch on a light bulb, the cold filament takes a big surge of current the instant before incandescence can drastically increase its resistance. The resistance ratio is approximately 10 to one.

But Ohm didn't learn that certain metal-like solids act in an opposite manner. (As insulators also do.) In 1833, Michael Faraday noted that *silver sulphide*'s resistance *decreases* with a rise in temperature. Ohm also was ignorant of the fact that the resistance of certain solids can be lowered by means of light.

In 1873, Britisher Willoughby-Smith discovered, quite by accident, that he could increase the current flow through a piece of the element selenium simply by shining a light on it, and that the brighter the light, the greater the increase. The pioneer television inventors used selenium to convert the output of their mechanical scanner to the video frequency.

The next step was taken by Antoine Henry Becquerel, who discovered that light can even *generate* electricity. For example, light falling on a layer of copper oxide with a copper backing will generate a small potential. This type of generator "powers" the photographer's light meter. Finally, the resistance of these solids is much greater in one direction than in the opposite direction.

The materials we have been talking about are called *semiconductors*. We can describe them much more quickly usin' the technical terms. The semiconductor is characterized by (1) a negative temperature coefficient of resistance, (2) photoconductivity, (3) photoelectromotive or photovoltaic force, and (4) rectification. This is quite a passel of virtues for a material that is neither a good conductor nor a good insulator. No wonder semiconductors have, within the past decade especially, created a revolution in electronics, chiefly through the development of the little *transistor*.

Some semiconductors other than those mentioned above are carbon, galena (lead sulphide), carborundum (carbon and silicon), silicon, and germanium. The most important of these is germanium, the element most widely used in transistors. The problem of using silicon in transistors was solved by Texas Instruments. They function at much higher temperatures than the germanium type.

Note that three of these *crystals,* carborundum, galena, and silicon, were used as detectors in the early crystal sets, the latter two with the cat-whisker contact. As we know, detection is a matter of rectification. Since they were snatched from their lodgings in the Smithsonian Institution, these crystals have left a vacancy in that museum of antiquity that, in time, might be filled by the pentode. The resurrected crystal, in the form of the transistor, is fast replacing the tube in many different areas of electronics.

The idea of the transistor can be attributed to William Shockley, solid-state physicist with the Bell Telephone Laboratories. In the conventional tube, the negative electrons from a hot cathode are controlled, on their way to the positive plate, by a voltage on the grid of fine wires through which the electron stream must pass. As the grid's electric (electrostatic) field exercises more control over the stream than the plate's field, the result is amplification. (See pages 75–76.) Dr. Shockley asked himself if it were not possible to obtain amplification by using electric fields on the relatively few free electrons inside a *semiconductor*.

The first transistor, like the crystal detector, used cat-whisker contacts—two of them. It was built by J. Bardeen and W. H. Brattain, also of the Bell Laboratories. Together with Shockley, they were awarded a Nobel Prize in physics for their achievement. The movement of the electrons through one of the fine wires touching the tiny piece of germanium controlled the movement through the other point effectively enough for amplification.

The point-contact transistor made its bow to the electronic

world in June, 1948. Although the gain was small, the power output low, frequency range limited, and the noise abundant, the point-contact transistor's pygmy proportions, meager power requirements as compared with the tube, ruggedness, and predicted long life, made intensive development certain.

Three years later, a more versatile version appeared, the *junction transistor,* conceived by Dr. Shockley and made by M. Sparks. Later, J. N. Shive, another Bell scientist, invented a germanium *photo-transistor* unit. Today, at least 30 companies in this country are engaged in transistor research and manufacture.

In their pure state, and at ordinary temperatures, crystallized semiconductors, such as germanium and silicon, are not semiconductors at all. They're wonderful insulators. The catch is that, unlike the driven snow, we don't find *pure* germanium or silicon crystals lying about.

In Chapter 2, we touched upon the theory of the atom, whose electrons are confined to shells, the shell nearest to the nucleus having a capacity of only two electrons, the next one eight, et cetera. Eight of copper's total of 29 electrons are lodged in the second shell and 18 in the third, which leaves a single electron in the fourth shell free to travel. Germanium's atomic number is 32, which means it has four electrons in its partly filled fourth shell, whose capacity is either eight or 18. Silicon, whose atomic number is 14, also has four outer-shell electrons.

Logically, this would seem to indicate that since germanium (and silicon), have four times as many outer shell electrons as copper, they should be four times more efficient as conductors. However, germanium's four outer electrons are far from free. They form links with the four electrons in adjacent atoms, binding them together. Each atom accomplishes four such bonds with other atoms, creating an electrically-stable five-atom unit or molecule, of which the crystal is composed. As we pointed out in Chapter 2, the combining electrons are called valence electrons.

Impure germanium is in the semiconductor class since some of its electrons are left over from the crystallizing process. Used for transistors, germanium first must be highly purified so that *selected* impurities can be added. The process entails pulling up the germanium in a crucible from molten material. As it cools, it solidifies in the form of a crystalline rod. At some point during the purifying process, an impurity may be added or, it may be added later.

Suppose, for example, a tiny amount of arsenic is added to the molten material. Arsenic's atomic number is 33, which gives it one more electron than germanium. In other words, its outer shell has five electrons instead of four, as germanium does. So when the arsenic atom combines with germanium atoms in the lattice structure of the crystal, a single electron is left over. These free electrons, one from each of the arsenic atoms that have been added, convert virtually pure germanium from an insulator to a semiconductor.

Because the conductivity is due to a surplus of free-to-travel negative electrons, this germanium is called the *n*-type (*n* for negative). Phosphorus and antimony also may be used for *doping* *n*-type germanium. In addition, the transistor requires another germanium, the *p*-type (*p* for positive), which is obtained by doping with boron (5), aluminum (13), or gallium (31).

Gallium, with one less electron than germanium, has three electrons in its outer shell. So when a gallium atom replaces its germanium counterpart in the crystalline structure, a vacancy occurs—a *hole* in each atom. This hole is, in effect, a positive charge. When the holes move through the germanium, they constitute a current flow, as does a movement of negative electrons.

How does one move a hole? If holes are movable they can be collected, which means that a man should be able to gather up enough old post holes for a cheap swimming pool in his back yard. As a matter of fact, for holes to move in one direction, electrons, taken from a nearby neutral molecule, must move in the opposite direction. As the electron fills a hole, it leaves another hole in its wake, which, in turn, is filled by another electron. However, the practical effect is that of a movement of positive charges. And the two types of germanium (or silicon), opposite in polarity, when used in *combination* make the transistor possible.

Materials such as arsenic, phosphorus, and antimony, which provide the semiconductor with electrons for conduction, are called *n*-type impurities, or *donors*. Boron, aluminum, and gallium, which remove electrons from the semiconductor atoms to create holes, are called *p*-type impurities, or *acceptors*.

If you place *any* two conductors back to back, and apply a voltage between them, there usually will be some difference in resistance with a change in polarity, though it may be slight. In other words, more current will flow through the junction of the two conductors in one direction than will flow through in the opposite direction, and this process is called rectification. Thus the junction can function as a diode to rectify an alternating current.

A semiconductor alone often makes a good rectifier, as in the case of the old galena detector. But a semiconductor in juxtaposition with another material can give us a better rectifier. A selenium rectifier, for example, consists of a thin coating of selenium on a steel, aluminum, or nickel-plated iron base plate, with an alloy of some sort placed on top for a contact plate. With the positive potential connected to the base plate, the negative to the top plate, this device offers very little resistance to an applied voltage. Reverse the connections, however, and the resistance will be very high.

Another commonly-used *barrier* material for making a solid rectifier, or diode, is copper oxide. The development of *n*- and *p*-type semiconductors has made possible much more efficient diodes for rectification. Germanium and silicon diodes are used in vast quantities throughout the electronics industry today, especially in computers. They are also replacing the old selenium diode in radio receivers.

Fig. 41 illustrates an *n-p* germanium junction diode. First, let's consider it without the battery connection. The *n*-type germanium has a surplus of electrons, left over from the substitution of foreign atoms for some of the germanium atoms in the crystal latticework. The added impurity in the *p*-type germanium has created an opposite condition: the combining process has made holes in the latticework that are, in effect, positive charges. At ordinary temperatures, neither the electrons nor the holes move very far.

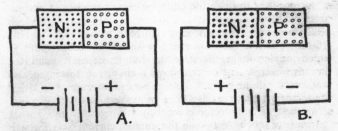

Fig. 41. An *n-p* junction diode rectifier. (A) Negative against *n*-type germanium and positive against *p*-type germanium permits current flow. (B) With the potentials reversed, the current flow is extremely small.

Now, let us apply the voltage as indicated in the diagram—positive against positive, negative against negative. The positive charge moves the positive holes toward the left, into the junction; the negative charge moves the negative electrons toward the right, into the junction, where the two opposite charges can combine. Of course, in order that the holes can move toward the left, the electrons must fill the holes, leaving other holes behind them, as we explained before.

Reverse the potentials, as illustrated at B. With negative against positive, and positive against negative, there is scarcely any movement of current through the *n-p* junction. This is because the negative battery voltage pulls away holes from the *p*-type material at the same time that the positive voltage pulls away electrons from the *n*-type material. (Opposite charges attract.) This leaves the barrier high and dry; in effect, it has become an insulator.

Now we come to the junction transistor, which can be made to amplify and oscillate like a tube. A photo transistor unit even responds to changes in light, such as occur in a phototube. Fig. 42 shows an *n-p-n* transistor. A sliver of *p*-material, less than one-thousandth of an inch thick, is sandwiched between two much thicker sections of *n*-material.

The extremely-thin *p*-material is called the *base*. The input is called the *emitter,* the output the *collector.* The symbol of the transistor is shown at B.

The battery on the emitter side, negative against negative, moves the *n*-material's electrons across the *junction* into the base. The collector's side is the same as in Fig. 41B, negative against *p*-material, and positive against *n*-material. No current should flow. Yet since the emitter side has moved excess electrons into the base, the situation has changed, and we have a current flow through the collector circuit to give us an output.

Suppose we *vary* the input current between the *n*-type emitter and the base by using a signal in series with the steady bias voltage. Because the output depends upon the input's electrons for current flow, a change in the input also varies the output current. And if an output current is larger than the input current, we have *amplification*.

Actually, the collector current may not exceed the emitter current by much. Yet since the emitter current acts through a low resistance, and the collector current through a high resistance, its *power* output will be greater. ($P = EI$. $E = IR$. Substituting IR for E in $P = EI$, we get $P = I^2R$.) In the

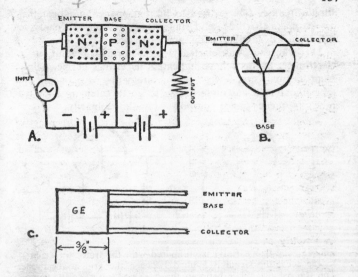

Fig. 42. (A) An *n-p-n* transistor, in which the current flows through the emitter and base, produces an output current through the collector and base. Output power can be considerably greater than input power, and this means amplification. (B) is the symbol for transistor. (C) presents a side view of one of the tiny transistors enclosed in a protective plastic case, ready for installation.

p-n-p transistor, the output consists of a current flow of "holes" instead of electrons, but the principle is the same.

When comparing the transistor to the triode, think of the emitter as the cathode because it furnishes the supply of electrons; think of the base as the grid because it controls the electron flow; and try to imagine the collector as the plate because it has the output current (Fig. 43). As in the case of the triode, the output (collector) power is greater than the signal power. The transistor also will oscillate through feedback, either inductive or capacitive.

Transistors can be used in both radio and audio frequency circuits; they work well as oscillators, clippers, and in many other ways. However, you can't substitute the transistor for the tube in a tube circuit. It not only requires much lower voltages, but different values of impedance between the circuits for matching. Countless magazines and scores of books illustrate practical transistor circuits for all purposes. In this simple treatise, I have tried to present enough practical theory to make the little solid stage device less of a mystery to those already familiar with tube operation.

The transistor has started a slow-burning revolution in electronics. But as I have indicated, its virtues derive mainly from its small size, low power requirements, ruggedness, and long life. It still has quite a way to go to catch up with the tube's frequency range and power-handling capacity.

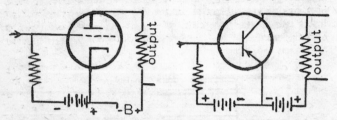

FIG. 43. The triode (left) is compared with the transistor (right). In the tube, the signal is applied between the grid and cathode. The transistor input is between the base and emitter. The output is taken from between the collector and emitter. However, there are several other ways of connecting the transistor so the output power is greater than the input power.

Transistors first found their way into computers, missiles, satellites, hearing aids, and small portable radios. But the transistor hasn't taken over the missile and satellite field entirely. Tube engineers have been fighting back, and today tiny triodes operate in both transmitters and receivers of the compact, sub-miniature, low-power jobs required by these high-flying vehicles.

The use of transistors in TV sets and hi-fi amplifiers has come more slowly. Eventually, they should supplant the relatively inefficient (inefficient from a power standpoint) hot cathode vacuum tube in both fields. At least transistors should be the next step in streamlining these devices. The deciding factor for television should be a minimum of servicing, decrease in size and weight, and perhaps lower cost. For high fidelity the transistor must provide still higher fidelity, plus longer life with less deterioration caused by heat.

Whether the transistor amplifier really delivers higher fidelity, the so-called "cleaner sound," than a tube amplifier is a matter of controversy among sound engineers. Transistor advocates cite its absence of microphonics (sound from vibration of elements inside a vacuum tube, no argument here), lower hum level, smoother and wider frequency response, higher damping factor and improved transient response (both explained in the next chapter) and absence of distortion from an output transformer used with tubes. Let the buyer compare.

Engineers are divided over the relative merits of transistor and tube in the FM stereo tuner, now an integral part of most hi-fi rigs. The problem is in the front end. A transistor as linear and as sensitive as the tube is hard to find, and there is also the matter of matching the low impedance transistor to the tuner.

In fact, in all parts of the set, a big difficulty now is locating the suitable transistor for the job. Even when one with the proper characteristics is found, lack of uniformity may be a problem. Given two transistors of the same type, one is apt to have twice the gain of the other. Many are also priced out of reach at the present time. Despite the difficulties, however, most high fidelity manufacturers now have transistorized hi-fi equipment on the market, including Trans-Tronics, Heath, Lafayette, Omega, Scott and Altec Lansing.

11

THE WHY OF HI-FI: THE AMPLIFIER

ONE summer afternoon, in the year 1912, a 39-year-old man of slender build and medium height, who sported a mustache, was seen acting very strangely on a street in Palo Alto, California. He was walking slowly along the sidewalk with his head bent forward in a manner indicating that both ears were cocked. He was listening to the faint sound of distant telegraph signals, which were coming from a horn in the window of a nearby laboratory.

The suspiciously-acting individual was a Yale Ph.D. named Lee de Forest, and he was testing the power of the first electronic amplifier of sound. As Dr. de Forest tells it in his autobiography, "I placed the loudspeaker in the laboratory window and walked down the street until the threshold of clear audibility was reached. After I had obtained a two-block gain, I felt reasonably satisfied with the Audion amplifier."

His "two-block amplifier" consisted of three of his little Audions, fed with telegraph signals from a wire recorder. He began by amplifying the audio frequencies because he was seeking a device that would remove the long-distance telephone call from the realm of high adventure. Three years later, the telephone company was demonstrating the new marvel of trans-continental telephony at the San Francisco World's Fair. Although de Forest's little Audion had made the invention possible, the company kept it secret. Maybe they were ashamed because they had paid him only $50,000 for a device that was worth so many millions to them.

Dr. Lee de Forest, "Father of Radio," and in fact, of the

whole electronics industry, died on July 1, 1961, at his Holly-wood home. He was 87.

We saw in Chapter 6 how de Forest developed the Audion for detecting dots-n-dashes. While it amplified the signal at the same time that it detected, de Forest didn't try to make it work as an amplifier until almost six years later. In Chapter 7, I told the story of how this Audion-amplifier-oscillator made radio broadcasting possible. This tube amplifier is still the cornerstone of high fidelity or was until transistors entered the picture recently.

In the same year that de Forest first hooked up several of his Audions in tandem, or in *cascade,* to use the technical term, for his "two-block amplifier," I was one of those who were greatly impressed by the naturalness of the voice of the great Negro comic, Bert Williams, as recorded on an Edison Record. When Williams would plaintively inquire concerning his involvement in a train wreck, "Who took that engine off my neck?" and would answer himself with a gloomily sarcastic "Nobody," it seemed as though he were right in the room with me.

The mechanical phonograph that evolved from Edison's tin foil device persisted until 1924. In that year, they began cutting records electronically. At the same time, Victor announced its revolutionary new Orthophonic line with the old mechanical reproducer. Hailed in the ads as the newest marvel of the age, it was made obsolete a few months later by Brunswick's Panatrope with its electromagnetic pickup (and amplifier) from General Electric.

As I have explained, this pickup is simply an electromagnetic generator. The phonograph needle, vibrating in the record's grooves, is attached to a magnetized piece of iron, which generates a current in a fixed coil, or coils; or the needle is attached to a coil that moves back and forth in a magnetic field to generate currents in itself. Without de Forest's tube to amplify them, these currents are too feeble to operate a loudspeaker.

De Forest, by the way, also was the first (1916) to use this electric pickup, a moving iron device. Although he always applied for a patent on any of his discoveries that had even the remotest chance of ultimately proving profitable, for some reason, he failed to drop this one into his bag, which eventually contained over 300 patents.

When the moving coil dynamic speaker was added to the electromagnetic pickup and tube amplifier, the groundwork was prepared for high fidelity. These devices, as I have indicated, greatly improved the quality of phonograph music during the twenties. Despite the depression, or perhaps because of it, the thirties proved to be radio's Golden Age. As a result, the phonograph made little progress, featuring cabinetry rather than quality, usually with a radio receiver built in.

Platters didn't have much to offer in competition to the goodies via kilocycle that included Amos 'n' Andy, Ed Wynn, Eddie Cantor, Mr. District Attorney, Myrt and Marge, Joe E. Brown, Al Jolson, Fred Allen, Charley McCarthy, Ma Perkins, Hit Parade, Jack Benny, Bing Crosby, and Elmer Blurt, the Low Pressure Salesman. FM, Major Armstrong's contribution to the hi-fi arsenal, didn't make its bow until 1941.

The term high fidelity came into being following World War II, when a few individuals, mostly engineers, began assembling their own home music systems. Since the components they used were far superior to those in the hand-rubbed mahogany cabinets, they presented a whole new vista of sound reproduction.

Three thousand sound bugs attended the first New York Audio Fair, held at the New Yorker Hotel, in October, 1949. But perhaps hi-fi's biggest boost of all came when the old 78-rpm shellac record was replaced by the long-playing, 33 1/3-rpm microgroove vinyl. Peter Goldmark, engineer-in-charge of Columbia Records, introduced the revolutionary new disc on June 18, 1948. With it was introduced the jewel stylus (sapphire or diamond), which made possible the light-weight tone arm greatly reducing record wear.

The cult that started among the handful of hi-fi-natics finally grew to such proportions that the manufacturers of commercial sets were moved to commit wholesale larceny. They stole the label for their inferior, hot-from-the-assembly-line, boxed-to-go products, from the $29.95 portable to the "de luxe" models. Recently, when another magic word appeared, the manufacturers were much quicker on the draw. Now everything they make is *stereo*.

Is the term high fidelity semantically sound? If we can have degrees of fidelity, why not degrees of roundness? When a man marries, should he select a bride who swears she will be *faithful* to him? Or would he be wiser to settle for one who only promises to be *highly faithful?*

The term, of course, was justified by hi-fi's need to grow. Each technical advance was hailed at the time as the utmost, the "mostest," in fidelity to the original. Consequently, the next rung up on the ladder had to be *higher* fidelity.

The first concern of the early audiophile was to build an amplifier that would pass all of the audio frequencies that appeared at the input with something resembling equal strength. The shellac records of those days had a rather limited frequency range, as did AM radio and the early tapes. This situation has improved greatly for both records and tapes, as well as by the introduction of FM radio with its wider frequency range and lower noise content.

The limited-frequency spread wasn't the fault of the tubes. They still give us sufficient trouble, but for other reasons, as we shall see. The villain was the coupling device between the tubes.

In the radio receiver, the couplings between the early stages, before the detector, usually are tuned. And if the circuit tunes *broadly* enough, few of the audio frequencies that are part of the tuned-in frequency are lost or diminished. After the detector, the coupling stages for the audio frequency alone present an entirely different problem, and, of course the hi-fi amplifier is concerned with audio frequencies only.

De Forest's original two-block audio amplifier used *transformer coupling* between the tubes. For the low frequencies in the audio range, transformers with iron cores, as shown in Fig. 22, page 90, are required. (The iron core greatly increases a coil's reactance, making fewer turns necessary.) The ratio of turns in his audio transformers was one to one; therefore, there was no voltage increase, as there would have been if there had been more turns in the secondary than in the primary coil.

In the early days of radio telegraphy, we took advantage of turns-ratio for greater amplification. But an audio frequency transformer has a resonant point, determined by the number of turns and the quality of the magnetic core. It passes the resonant frequency, and those close to it, much better than the others.

When we plot this characteristic on paper, we obtain a resonant peak, which enables us to say that the transformer *peaks* at a certain frequency. This was fine for the dot-n-dash signals. The old spark sets were on 500 kilocycles (600 meters), and an audio transformer that peaked near this frequency was exactly what the doctor (de Forest) ordered.

Unfortunately, during the early years of broadcasting, this same old transformer was handed down to the manufacturers of the first tube sets. Even if all the audios that went into the amplifier managed to show up in the output, those frequencies on both sides of the resonant peak were greatly diminished. The entire audio spectrum—at *least* between 100 and 5,000 cycles—should reach the loudspeaker with something close to equal strength. Another signal change that occurs in an audio transformer results in *intermodulation distortion*, which we shall treat later in considerable detail.

The audio transformer has been enormously improved. By placing the resonant peak outside of the audio range—say at 70,000 cps.—the amplification of the audio spectrum can be kept fairly uniform. Good engineering, together with recently-developed superior core materials, also can drastically reduce distortion.

Then why aren't we still using them for hi-fi? As a matter of fact, we are, though mostly between the final pair of amplifying tubes and the loudspeaker. This is the *output transformer,* whose principal function comes under the head of *impedance match*.

The high impedance of the plate circuit, in the thousands of ohms, must be matched to the low impedance of the voice coil, usually four, eight, or 16 ohms. But the other audio stages of today's hi-fi rigs are almost universally coupled by an R-C network (Fig. 27; page 103). The change-over took place largely as a result of the cost factor. A transformer is expensive, while you can purchase a couple of resistors and a small capacitor for less than a dollar. Transformers also present a hum problem.

Obviously, an R-C coupler doesn't furnish any voltage increase. In fact, the output voltage always is less than is indicated by the tube's amplification factor. With improved tubes, notably the pentode, and the use of a final push-pull stage, voltage amplification between the tubes was no longer necessary. As we learned in Chapter 8, a little .01 µfd capacitor in the R-C network, which is necessary to prevent the high plate voltage from encroaching upon the grid of the next tube, has its favorites among the frequencies. It slights the lower audios, those under 250 cycles. Only the middle frequencies, 250 to 5,000 cycles, pass through unscathed.

All of the frequencies above 5,000 cycles tend to be drained off by what is called "stray coupling," which consists of capacitances between elements in the tubes, such as the grid and cathode, and between the conductors and the ground. Small as they are, these capacitances offer little enough re-

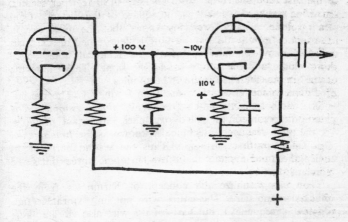

Fig. 44. Direct coupling between audio stages.

sistance to the higher audios to be taken seriously, though their problem isn't as dismal as our analysis might indicate. You can reduce frequency favoritism to negligible proportions by selecting the correct values of capacitor and resistors for the type of tube you use. This information may be found in the tube manuals.

Some purists insist upon *direct* coupling between audio frequency stages (Fig. 44). The grid uses the same high positive voltage as the plate of the preceding tube. It is then made negative in relation to the cathode by placing an even higher positive voltage on the cathode. For example, a 100-volt positive grid voltage will be 10 volts negative with respect to a 110-volt positive cathode. *Less* positive always means *more* negative. The grid is grounded as shown in the figure.

An R-C network not only affects frequency response; it also has a bearing upon distortion in the tube. But before we discuss this we should first briefly review the theory of sound.

Harmonics! This is the key word to a thorough understanding of the problem of sound reproduction. That sound consists of waves in the *air* has been known ever since 1680, when Robert Boyle demonstrated that a clock ticking in a vacuum is as silent as a sawmill on Sunday. And even slow-learners in the fifth grade should know that sound waves consist of alternate compressions and rarefactions of the air, which travel outward at approximately 1130 feet per second, growing weaker inversely as the square of the distance from their source. Finally, everyone knows that the difference in the sound from a fiddle and a flute is determined on the basis of frequency. Then why doesn't the identical note played on a fiddle and on a flute *sound* the same? They share the same *frequency*. The reason is harmonics.

Pitch is a better word to use than frequency. The musical note you hear is the pitch frequency, *plus harmonics*. The harmonics, sometimes called overtones, are exact multiples of the pitch frequency, technically known as the *fundamental* frequency, or first harmonic. Thus the second harmonic is twice the fundamental, the third harmonic three times the fundamental, et cetera.

You can visualize the concept of harmonics from the plucked piano wire. The entire wire not only vibrates (fundamental frequency), but parts of the wire also vibrate along fractions of its length. For example, the two half sections vibrate at double the fundamental to give us the second harmonic.

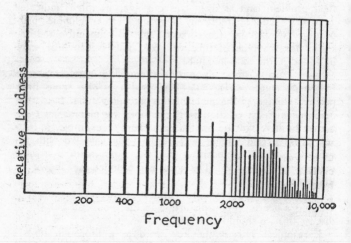

FIG. 45. Plot of the relative loudness of the various harmonics in G below middle C, when played on the violin. The fundamental frequency is 196 cycles.

The stringed instruments, including the piano, are richest in harmonics. A violin note may have as many as 20 harmonics that are measurable. The flute and the piccolo come closest to producing pure tones. Some notes on the flute may have only a single harmonic of any intensity.

To add to the complexity, the number and amplitude of the harmonics of any single note can change with intensity. This explains, in part, the skilled violinist's extremely subtle control of his instrument. He can vary the flavor of a single note by moving it up or down in the intensity scale to suit his interpretation. The pianist also.

Fig. 45 reveals the relative loudness of the measurable harmonics of G, below middle C, whose fundamental is 196 cycles, when played on the violin. Note that in this instance, some of the harmonics possess greater volume than the fundamental. In most cases, however, the fundamental is the strongest.

Fig. 46 pictures this G as seen on the oscilloscope. At the top is the combination of the fundamental note with its harmonics; their relative loudness is shown below.

Very few instruments have a fundamental note that transcends 2,000 cycles. These include the pipe organ (8,000 cycles), piano (4,000 cycles), harp (2,500 cycles), violin (3,200 cycles), flute (2,500 cycles), and piccolo (3,800 cycles). Percussion instruments, such as the drums and cymbals, may have higher fundamentals—the cymbals around 16,000.

If, in recording an orchestra, all but the fundamental notes of the instruments were filtered out, the result would be rather anemic. To differentiate properly between them, usually their first four harmonics are necessary. On the other hand, to impress on record or tape *all* of the harmonics from an orchestra would be impossible. Even if it were possible, it would be labor lost because few of us have ears sharp enough to hear anything above 15,000 cycles, and most of us, especially if we're advanced in years, fall far short of this dizzy height. Bats, dogs, and cats do much better, but they represent such a minute portion of the hi-fi market that manufacturers haven't seen fit to consider them. (Or have they?)

Most makers of hi-fi amplifiers seek to impress the buyer with graphs indicating that their machines will pass all of the frequencies from 30 to 15,000 cycles with favoritism toward none—*flat* from 30 to 15,000 cycles is their boast. And a really good amplifier will approach this closely enough for happy results, though whether the *speaker* will complete the job by satisfactorily converting them to sound is another ques-

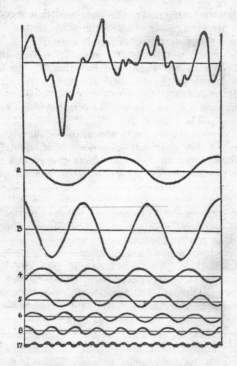

Fig. 46. How the 196-cycle violin note looks on an oscilloscope. Its principal harmonics or *dominant partials* are shown below.

tion. Yet we see advertisements boasting of amplifiers that are flat from 20 to 50,000!

Even though we don't consciously hear the super-15,000 harmonics, the mystics tell us that the music is improved; they can even prove it by listening tests. First, they play a record through the 50,000-cycle amplifier. Sounds real good. They add a filter that chops off all frequencies *above* 15,000. Sounds not so good. However, engineer Norman H. Crowhurst, prolific writer on high fidelity, claims that the filter is responsible; the filter adds *transient distortion*, which accounts for the difference. So maybe manufacturers of the U-2 amplifiers are making them for the birds and dogs after all.

How about very low frequencies, those under 100 cycles? A frequency below 50 cycles is difficult to reproduce at a good volume level. You may *think* you're hearing it when actually most, or all, of its energy is in its second harmonic, 100 cycles. The same condition applies to still lower notes. A note as low as 20 cycles can hardly be called music; it's more like a noise we feel rather than hear. The lows are also more vulnerable to distortion in the speaker.

Another interesting fact about the lows is that, given a low note and a high note of equal intensity, our ears respond best to the high note. Starting with 700 cycles, and dropping steeply off to the least audible frequency, our ears are progressively less sensitive. This effect is greatest when the *intensity* of the two notes is low. At high intensities, of around 90 decibels, such as a symphony orchestra in a concert hall can produce, we hear both notes almost equally well. The high notes (above 4,000) also drop some in relative intensity, though not nearly as much as the lows. However, concert hall volume doesn't level off the highs to any extent.

This is why some audiophiles seek to improve upon de Forest's two-block amplifier with three- and four-block jobs or better. When the volume control is turned way up, they don't lose as many of the lows, though they may lose plenty of friends in the neighborhood. Tone control circuits can help to compensate for our hearing deficiencies.

A moment ago, I mentioned decibels, the unit we use to measure the differences in sound intensity. The decibel is one-tenth of a *bel,* after Alexander Graham Bell, teacher of the deaf, who was the first man to obtain a telephone patent in this country. However, the decibel is usually referred to by its abbreviation, db.

The decibel is not an expression of true sound intensity. It merely reveals the noticeable *difference* in loudness between two sounds. When we say of an amplifier that it is flat, plus or minus 3 db, between 30 and 15,000 cycles, we mean that there is no difference in loudness over this frequency range that exceeds 3 db. Let's see how much of a difference this is.

Suppose an amplifier is rated at 25 watts. We double its output to 50 watts. If we were to pick up these sounds with a microphone, and measure the result, the sound power of the 50-watt amplifier would be twice as great as the 25-watter. But do our ears tell us this? Do 50 watts *sound* twice as loud as 25 watts?

The answer is no. The difference between the two amplifiers is barely perceptible—three db. However, the difference is more pronounced at high volume—another reason why some audiophiles like to turn their sets way up. Contrasts between the program's volume levels are more evident; they pertain to what is called *dynamic range*, which is very important to high fidelity.

When the power output is multiplied by four, the increase on the decibel scale is six db. Eight times as loud means nine db, 10 times as loud, 10 db.

So we see that our hearing apparatus makes the decibel scale, if not necessary, at least convenient. This scale is approximately the same as the logarithmic scale, while sound intensity uses simple arithmetic. The faintest perceptible sound, called the "threshold of hearing," may be considered as a reference level for zero db. An intensity one million times greater is equal to 60 db. Listening levels for home music systems vary between 40 and 80 db.

To convert watts of power increase to decibels requires some familiarity with arithmetic and a table of logarithms. The formula is db $= 10 \log_{10} P_1/P_2$, the P standing for power. If the input power to an amplifier is .05 watts, and the output power is 10 watts, then $10/.05 = 200$. The $\log_{10}$ of 200 is 2.3010; and $10 \times 2.3010 = 23$ db.

In our story of the superhet, we saw how the heterodyning or "beating together" of two frequencies creates a number of new frequencies. Both radio and TV sets mix the incoming carrier frequency with a local oscillator frequency. From this mixing arises a *difference* frequency, called the intermediate frequency (i-f), which is the same for all stations tuned in. A fixed-tuned i-f circuit rejects all the other products of the mixing. These other products include not only the *sum* of the two or original frequencies, but also a number of sum and difference frequencies of the harmonics.

Now you may have wondered why the complex mixture of *audio* frequencies, including both the fundamentals and a great many harmonics, don't mix with each other to create new frequencies. First, we must realize that in radio and TV receivers, the incoming carrier and the local frequency wouldn't mix if the mixer tube were linear, and both frequencies went to the same grid. The mixing is achieved through the tube's non-linearity and the use of separate grids.

In the case of the audios, they all go to the same amplifier grid. At the same time, we realize that no tube is completely linear. There always is a certain amount of mixing, of heterodyning, among the audios. And the distortion that results is usually much more distressing than the mere loss of some of the frequencies at the two extremes of the audio spectrum.

Non-linearity causes two kinds of distortion among the audios—harmonic and intermodulation. Before we go any further into distortion, it would be well to review the matter of tube non-linearity, which causes it.

In Chapter 8, we discussed the characteristic curve. (See Fig. 19; page 82.) This curve tells us that for minimum distortion of the wave form, we must select a grid bias that will place the signal on the "straight" portion of the curve, and not overload the amplifier. However, the so-called "straight" or "linear" portion of the curve never is completely so. Thus, the rise and fall of the current in the plate circuit, in its amplified form, never follows exactly the rise and fall of the input voltage on the grid.

Here is a way to figure some of the non-linearity of an amplifier. Each half cycle of the grid voltage produces a different plate voltage swing. For example, a two-volt *positive* grid may cause a 50-volt plate swing, whereas the *negative* half of this cycle may produce a 54-volt swing. The difference is four volts. Incidentally, the difference is always less when the grid voltage is lower; overloading always increases distortion.

If we compare the output of a voltage amplifier with its sine wave input, as we can by either drawing them on paper, or projecting them on an oscilloscope screen, we can actually *see* the distortion. But what does this distorted sine wave have to do with *harmonics?*

For the last time, we'll go back to one of the Greeks. A great mathematician named Eudoxus, a friend of Plato's, started it all when he suggested that an irregular curve was composed of regular components. You might read that sentence twice, for it anticipates *harmonic analysis,* one of the greatest discoveries of modern mathematical physics. The discoverer, Jean Baptiste Joseph Fourier (1768-1830), developed mathematical equations that make it possible to separate the component harmonics from any motion. Today we can do it electronically with a *harmonic analyzer*.

The proof of the pudding is in the listening. Not only can the harmonics be measured, but the sine waves can be put together to create any sound you wish, from plucked guitar string to church bell. If they were combined in the quantities and phase relationships indicated in Fig. 46 (page 148), the sine waves would give us the violin tone whose graph appears on the top line. A single sine wave alone produces a note that resembles a flute played softly.

It should be clear by now that non-linearity makes amplification of the complex mixture of notes from an orchestra something of a problem. For the tube's non-linearity adds some harmonics of its own. From this phenomenon comes the term *harmonic distortion* or *amplitude distortion*. They mean the same thing because, as Eudoxus suggested, a change in the wave's amplitude (shape) can be attributed to the harmonics.

The additional harmonics don't cause too much trouble because of their direct relationship to the original frequencies. The ear tolerates much of the second harmonic, less of the higher, even harmonics and still less of the odd harmonics. The real trouble results from the harmonics heterodyning (modulating) one another, to produce new frequencies, many of which are not multiples or sub-multiples of the originals. This is called *intermodulation distortion* (IM), and it is largely this distortion that adds harshness, rattling, grating, tinniness, even a blurred effect to the music. A decade ago, it was considered ideal if intermodulation distortion was kept under two per cent. Today, negative feedback can keep down distortion to a fraction of one per cent.

Distortion in an amplifier can be measured by feeding it with one or two sine waves, and connecting the output to a harmonic analyzer. Filter circuits in the analyzer separate all the sum and difference frequencies. As many as 20 distortion components from the two frequencies may be present in a sufficient amount to register on a meter.

Let's return to the audio amplifier with the R-C network. We'll start with the triode amplifier, connected as indicated in Fig. 47A. R_P represents the tube's internal, or plate, resistance. R_L is the load resistor, C the coupling capacitor, and R_G the grid resistor.

Fig. 47B is the *equivalent* circuit for Fig. 47A. As this diagram indicates, the tube operates as a voltage generator. The size of the generated voltage is equal to the input voltage (grid voltage swing), multiplied by the tube's amplification factor, less the loss in its internal resistance (R_P).

Since R_P and the load resistor, R_L, are in series, the voltage divides between them in direct proportion to their respective resistances. If we make R_L large in comparison with R_P, *most* of the voltage will develop across it, though some

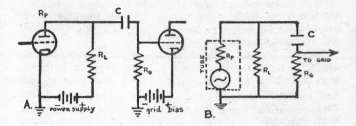

FIG. 47. (A) First stage of R-C coupled audio amplification. (B) Equivalent circuit for A. B reveals that since the internal resistance of the tube generator (R_P) is in series with the load resistor (R_L), the voltage divides between them. C and R_G are in parallel with R_L. The resistance of two circuits in parallel is always less than that of either circuit alone; therefore, C and R_G will reduce the tube's load resistance. If R_G is greater in relation to R_L, the total resistance is not critically lowered. Since the grid voltage is taken from across R_G, whatever resistance C offers to the signal is lost. However, C's opposition is only appreciable for frequencies under 250 cycles.

will always be lost in R_P.

We can't make R_L too large, or it will reduce the current flow through the tube to the point where it won't function properly. We can raise the plate voltage to compensate for this, but there is usually a limit to the amount of plate voltage available. And even if there were no limit, a tube, like a horse, can carry only so much of a load (volts times amperes) without overheating. Also, making R_L more than five times R_P doesn't help much as far as voltage gain is concerned. Keeping distortion low is more important, and is also an important consideration in determining the value of R_L, as we shall see later.

It is clear by now why the actual voltage gain in one of these amplifiers is always less than the amplification factor. However, a number of other statements about the audio amplifier still require proof. For instance, in the chapter on TV, we said that a signal emerges from an amplifier inverted, "turned over," the positive half cycle made negative, the negative half cycle made positive. And we have indicated that the selection of components for the R-C network also has a bearing upon linearity, which determines the amount of harmonic and intermodulation distortion.

Fig. 48 is a graphic representation of what happens in a

6J5, a triode voltage amplifier, under operating conditions. Although the subject is advanced for this brief survey of electronics, we should be able to master it with a little high-voltage concentration.

The characteristic curve of Fig. 19 (page 82) plots the values of plate current (I_P) against the different values of grid voltage (E_G). Its principal use is to reveal the optimum grid bias for working on the straightest portion of the curve. Our present chart goes much further, and reveals changes in both plate current (I_P) and plate voltage (E_P) when a *changing* signal voltage appears on the grid. From this chart, we can determine, among other things, the amount of voltage amplification that actually occurs.

To illustrate all three values and their relationships, plate *voltage* as well as plate current and grid voltage, we must draw a separate curve for each one of a number of grid bias voltages. Then we can see at a glance how much plate current flows for any given plate voltage—with any of the chosen bias voltages on the grid. For a grid bias of —8 volts, a plate voltage of 230 will move six ma. of current through the tube.

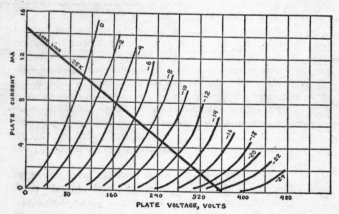

FIG. 48. A family of characteristic curves, each one showing plate current (I_P) and plate voltage (E_P) for a particular grid bias voltage (E_G). The load line is drawn from the point of maximum d-c plate voltage to maximum d-c plate current. By glancing along the load line, you can see the extent to which a change in grid voltage affects the plate voltage and plate current. Changing the value of the load resistor or the plate voltage, or both, changes the slope of the load line. The slope also reveals the amount of distortion in the amplifying stage.

With only 110 volts on the plate of this 6J5, the −8 grid voltage will prevent any current flow. This is the cut-off point.

How will the tube function as a voltage amplifier in an R-C circuit with a load resistor (R_L) in the plate circuit (Fig. 47)? The answer is found in the *load line*. The *slope* of the load line reveals both the actual *gain* of the tube, and the amount of harmonic distortion. And the slope is determined by the value of R_L and the supply voltage chosen. Fig. 48 depicts the load line for a 25,000-ohm resistor and a supply voltage of 360.

It is obtained in the following way. The lower end of our load line begins at the chosen 360 volts and zero plate current. It's just as if R_L were short-circuited and the whole 360 volts were on the tube's plate. On the other hand, if the *tube* were short-circuited, the entire 360 volts would be projected across the load resistor. The 360 volts across 25,000 ohms would result in a current flow of 14.4 milliamperes (ma.). (I = ER. I = 360/25,000. I = .0144 amperes, or 14.4 milliamperes.)

Between these theoretical extremes, the signal voltage on the grid varies the plate current through the resistance of the tube (R_P) and the load resistor (R_L) in series. Thus we have two voltages, one across R_L and one across R_P.

Note that the 25,000-ohm load line is absolutely straight. Therefore, R_L's voltage must be *linear*. Any change in current through the load line always causes a proportional change in voltage across it. This isn't always true of a resistor, though for all practical purposes, it must apply to the resistor used in this example. Look at the grid voltage change versus plate voltage change. In the case of the tube, Nature throws us nothing but curves. If the E_G lines also were straight, like the load line, and evenly-distanced, there would be no distortion.

Our supply voltage is 360, and the load resistor is 25,000 ohms. Suppose we select −8 for the grid *bias* voltage. The grid line for −8 volts crosses the load line at a point directly opposite a plate current of close to 5.4 ma. Dropping straight down from this point, we see that the voltage across the tube is 225. Thus with no signal on the grid, we have a 5.4 ma. plate current and a 225 plate voltage. As the total voltage of 360 must divide between the tube and resistor, the latter must have 135 volts across it. (360−225 = 135.) Ohm's law gives us the same figure. (E = IR. E = .0054 × 25,000. E = 135.)

Let's see what happens when we apply a signal to the —8 volt grid. For test purposes, we shall use the alternating voltage of a sine wave. This voltage becomes two volts positive and two volts negative once each cycle, which means a swing of four volts from peak to peak. The positive half cycle will reduce the grid's —8 volts to —6 volts; the negative half cycle will increase it to —10 volts. We now can determine the effect of a four-volt *grid voltage swing* on the plate voltage. It is this *plate voltage swing* that is passed on to the next tube.

Lines drawn straight down from the point of intersection, where the —6 and the —10 grid voltages cross the 25,000-ohm load line, reveal that the plate voltage changes from 190 volts to 225 volts. Thus, $255-190 = 65$ volts. As we obtain this 65-volt plate swing with a four-volt grid swing, the *gain* (A) of this stage must be approximately $65/4 = 16$. The amplification factor of the 6J5 is 20. The difference, as I have stated, is caused by the loss in the internal resistance of the tube. A bias resistor in the cathode circuit also will cause some loss, though not much since its value is small compared with the tube resistance.

We see that with the positive half cycle, which reduces bias voltage from —8 to —6 volts, the plate voltage goes *down*. While the negative half cycle, which raises the bias to —10 volts, causes the plate voltage to *rise*. This is our proof that an amplifier "turns over" the applied grid voltage, making the positive half cycle negative, and vice versa.

Why this inversion? This positive grid increases the voltage drop across the load resistor (R_L), and this drop is taken away from the plate voltage. On the other hand, the negative grid reduces the voltage drop, thereby increasing the plate voltage. In the above case, with the —8 bias voltage the voltage drop across R_L is 135 volts. ($360-225 = 135$.) The increased current flow, caused by the more positive grid, raises this voltage drop to 170 volts ($360-190 = 170$), while the lowered current flow from the more negative grid drops it to 105 volts ($360-255 = 105$).

Ohm's law will verify this. With —6 grid volts, the plate current is 6.8 ma. $E = IR$. $E = .0068 \times 25,000$. $E = 170$. The current flow times 25,000 for the —10-volt grid will equal 105 volts.

Our window into the mysterious workings of an amplifier also can reveal distortion. The —6-volt grid on the positive half cycle reduces the plate voltage from 225 to 190, a change of 35 volts. The —10-volt grid on the negative half

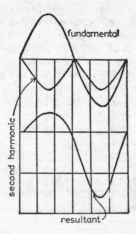

FIG. 49. A sine wave with second harmonic sine wave. The result is an asymmetrical wave.

cycle raises the plate voltage from 225 to 255, a change of 30 volts. This represents a five-volt difference between the rise of the negative half cycle and the rise of the positive half cycle. The sine wave that emerges from the amplifier will be different from the sine wave that entered it.

This kind of distortion, in which one half cycle is of a different amplitude from the other, is always *even harmonic distortion*—that is, second, fourth, sixth, et cetera. Fig. 49 illustrates the way in which addition of a second harmonic to the fundamental can throw the two half cycles out of balance. This unbalanced, or asymmetrical, distortion applies only to even harmonics. Odd harmonic distortion (third, fifth, seventh, et cetera) is symmetrical. (Fig. 50.)

Another glance into our window of Fig. 48 (page 154) should reveal the manner in which a different value of load resistor can provide more amplification and less distortion. A load resistor of 50,000 ohms might cross the curving grid voltage lines in such a way that the grid swing would indicate a greater gain and less of a second harmonic distortion.

I trust that we now understand why increased distortion results when we turn up the volume too high. The greater the grid voltage swing, the farther the plate voltage swing reaches that area where the grid voltage lines are progres-

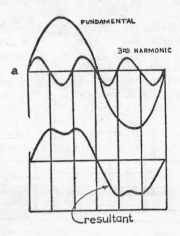

FIG. 50. A sine wave fundamental with a third harmonic sine wave added. The result (b) is a symmetrical distortion of the fundamental.

sively unequally distanced. *Too great* a swing will cause either one or both of the half cycles to be sliced off. This is known as *clipping*.

To simplify our explanation of voltage gain, inversion, and distortion in the voltage amplifier, we assumed that R_L alone determined the load resistance. However, Fig. 47 B (page 153) makes it clear that this isn't true. R_G, the grid resistor, must be taken into account. And the coupling capacitor C also affects the frequencies under 250, as we have pointed out.

Fig. 47B reveals that R_G is *parallel* with R_L. This means that the two resistors together draw more current than R_L alone, as two light bulbs in a floor lamp use more current than one alone. And if more current flows, the resistance must be less ($R = E/I$). So R_L's resistance is reduced by R_G's resistance in parallel with it. Let's see how this affects the selection of the value for R_G.

The simple equation for resistances in parallel is: $1/R$ total $= 1/R_1 + 1/R_2 + 1/R_3$, et cetera. Suppose, for example, R_L and R_G are equal. Then the total resistance would be half of R_L, or 12,500 ohms. ($1/R_T = 1/25,000 + 1/25,000.$ $1/R_T = 2/25,000.$ $2 R_T = 25,000.$ $R_T = -12,500.$)

To avoid reducing the value of R_L too much, R_G is made very large in comparison. For instance, if R_G has a value 20 times that of R_L, which is usual, then the total resistance is only a *little* less than that of R_L alone.

There is a simple formula for figuring voltage gain A of a triode from the amplification factor μ, the plate resistance R_P, and the total load resistance R_T. $A = \mu R_T/(R_P + R_T)$.

The 6J5's plate (internal) resistance is around 8,000 ohms. A grid resistor of 500,000 ohms can be used with our 25,000-ohm plate load resistor. These two resistors in parallel provide a total resistance of 23,810 ohms. Inserting a μ of 20, an R_T of 23,810, and a plate resistance of 8,000 in the above equation, we get, $A = 20 \times 23,800/8,000 + 23,810.$ $A = 15.$ This closely agrees with the gain of 16 we obtained from our graph.

Why didn't I give you this formula in the first place instead of asking you to take a long, hard look at the load line? I suppose I wanted to make certain I understood the theory myself. The blurb for one publisher's book on electronics for beginners boasts that it contains *no dull theory*. This reminds me of the chimpanzee that learned how to play the fiddle. *He* got by without bothering with theory, and even got a job on television, but I've always had a sentimental weakness in that direction.

A final word about the coupling capacitor C of Fig. 47 (page 153). It's in series with R_G, so that whatever reactance it has takes away some of the voltage across the impedance of R_G, which goes to the grid. With a value of from .01 to .1 μfd, however, this capacitor is practically a short circuit above 250 cycles. And remember that its reactance and R_G's resistance don't add arithmetically. To obtain the impedance across R_G, $Z = \sqrt{X^2_C + R^2_G}$, thanks to Pythagoras.

Many of the *aficionados* used to prefer low-μ triodes over pentodes in their hi-fi amplifiers because of their relatively low distortion; today, nearly everyone uses high-μ pentodes with negative feedback to reduce the distortion.

The pentode's family of voltage curves is different from the triode's (Fig. 51). The first thing that strikes us about these curves is that, unlike the triode, the plate current increases very slowly with increased plate voltage. This is be-

cause the constant voltage on the screen grid is doing most
of the pulling of the electrons away from the cathode space
charge for the plate current. By increasing the plate voltage,
you add few electrons to the stream heading toward the
plate. Of course, the signal grid operates as in the triode, and
compared with the plate's, its leverage on the electron stream
is much greater than in the triode—an advantage that gives
the pentode a much higher amplification factor. Special triodes can be built, however, with a very high amplification
factor.

The pentode also has a much higher internal resistance
(R_P) than the triode—approximately 25 times higher. For
if a change in the pentode plate voltage causes so little change
in the plate current, the internal resistance must be proportionately greater.

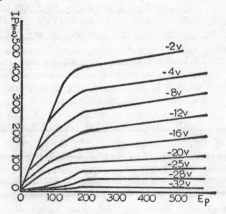

Fig. 51. A family of characteristic curves for a pentode power
tube. The lines are even more horizontal for the pentode voltage
amplifier.

The voltage gain in a pentode is directly proportional to
the value of R_T, as in the triode. But with the pentode's
higher internal resistance, it is impossible to use a load resistor many times greater, as you can with the triode. However, additional resistance is not necessary because of the
pentode's much higher amplification factor. The load resistor for a pentode varies from around 2,500 to 8,500 ohms,
depending upon the type of tube and the plate voltage.

Now, let's switch from the *voltage* amplifier to the *power* amplifier output stage. Fundamentally they're the same, except that the load resistance, which becomes an *impedance*, is selected on a different basis, and the amount of power handled is much greater. Whereas the output of a voltage amplifier may be only a few milliwatts, a power amplifier may be asked to deliver as many as 100 full-grown watts to the loudspeaker. Needless to say, it must be a much sturdier tube than the voltage amplifier.

An output circuit that makes any pretense at all to high fidelity uses a pair of tubes in what is called *push-pull*. E. H. Colpitts, the inventor of the oscillator circuit that uses capacitive feedback, also gave us push-pull. It was first used over 40 years ago, to increase the efficiency of the carbon microphone by permitting the addition of a second carbon button. The push-pull output amplifier provides approximately twice the power and cuts down distortion. The latter improvement is the most important, since additional power can be obtained equally well from two tubes connected in parallel. The conventional way to transfer power from amplifier to speaker is through a transformer. Why this is convenient will soon be evident.

Fig. 52 shows a pair of push-pull power tubes (triodes) connected to a speaker through an output transformer. A first glance at this ingenious hook-up may prove frustrating. When connected in series, if current is to circle through the tubes, one grid must always be positive while the other grid is negative, and vice versa. As a result, the plate current through one tube will always be rising while the other tube's plate current will be falling. And two equal currents, 180 degrees out of phase like this, will rub each other out. For they are directly opposed at all times.

The joker is the center tap on the transformer primary winding (P). This center tap causes the currents to the two plates to move in *opposite directions*. A change in direction is the same as a change from rise to fall and vice versa. Thus the plate currents, instead of cancelling, now rise and fall in phase.

The two tubes' combined current flow is approximately double the current from each tube alone. As shown graphically in Fig. 53, one half cycle seems to land directly on top of the other, though it's slightly more than difficult to pick a winner. This has a meaning for distortion.

With second harmonic distortion, one of the half cycles never rises quite as high as its companion. By combining a

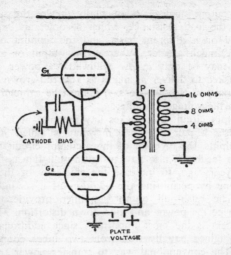

Fig. 52. Push-pull output stage of an audio amplifier that uses triodes. The primary of the output transformer is matched to the internal resistance of the tubes; the secondary is matched to the impedance of the speaker's voice coil. The transformer itself is a matching device (see text).

negative with a positive half cycle, as in Fig. 53, the inequality between them vanishes. They are averaged out. This is true for all even harmonic distortion, fourth, sixth, et cetera. All odd harmonic distortion escapes destruction.

In addition to this cancellation of even harmonic distortion, the superimposure of the two halves of each cycle in push-pull also irons out much noise and hum. It also provides other bonuses that the interested student may find in more technical treatises on the subject.

Since the power supply voltage is a direct current, it will encounter only the *resistance* of the wire in the transformer primary. As this is only a few ohms, the supply voltage that reaches the plate will not be radically reduced, as it is in an R-C-coupled voltage amplifier. Of course, when a signal is applied to the grids of the tubes, and a small part of the supply current starts *changing* (rising and falling), then the inductance of the transformer primary, with its voice coil load connected to the secondary, adds considerable reactance to the resistance. It's this variable reactance, plus the resistance, that provides the *load* for the audio frequencies, in

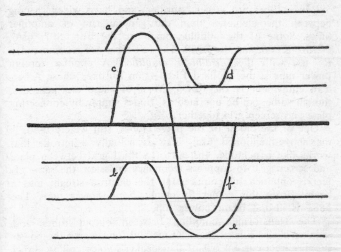

FIG. 53. Output current of a push-pull amplifier. The two half cycles, *a* and *b,* are added to make *c; e* and *d* become *f.* Thus, most of the asymmetrical distortion, somewhat exaggerated in the drawing, is eliminated.

contrast to the simple resistance (plus the capacitance of the little coupling capacitor) in the R-C voltage amplifier. The inductive reactance (X_L), plus the resistance (R), equals the impedance (Z). The addition is algebraic, as illustrated in the case of the coupling capacitor and grid resistor of Fig. 47 (page 153).

In the voltage amplifier, we use a load resistor with a high value relative to the tube's plate resistance in order to salvage as much of the division of voltage between them as possible. In this instance, we want *power*—volts times amperes—from the output of our amplifier. The tube is a generator, and over a century ago, Joseph Henry told us how to obtain maximum power from it: make the load resistance equal to the internal resistance of the generator. The *impedance* load of the transformer is no different. However, in matching the triode tube generator with the transformer, we should be certain that in the interest of lower distortion, the transformer impedance is approximately twice the tube impedance. Whenever the question arises of defeating distortion, or increasing power, the former always gets the nod.

Power tubes are larger, sturdier, and have wider spacing between the elements than the general run of amplifier tubes. Some of the popular types are the 6L6, 807, 6V6, 6CM6, KT66, and the 6550. The last three are recommended for use with the *ultralinear amplifier*. A popular foreign power tube is the Mullard EL84 from Europe, whose American equivalent is the 6BQ5. Most of these are pentodes, though some can be operated as triodes simply by connecting plate and screen grid together.

One of the oldest of the power tubes, still widely used, is the above-mentioned 6L6. This is actually a tetrode that works like a pentode. Instead of a third grid between plate and screen grid to suppress secondary emission, the same effect is obtained by compressing the electron stream into a beam. From this comes its name: *beam power tube.* The 5881 is also a beam power tube.

The transformer coupling between output tubes and speaker is also a very practical method for keeping the power supply direct current out of the speaker's voice coil. A direct current through this little coil would be disastrous since it would move it off center and cause it to heat up. When direct coupling is used, it must be protected by a large capacitor, one large enough to pass all the audios satisfactorily.

Now we come to the impedance match between the transformer and the speaker. The plate resistances of output power tubes are much lower than those of voltage amplifiers. For triodes, the internal resistance varies from 10,000 to 50,000 ohms; for pentodes from 15,000 to 100,000 ohms. The impedance of the little voice coil of the dynamic speaker is rated at four, eight, or 16 ohms in this country. Quite a difference. If the voice call were placed directly in the plate circuit, this gross mismatch would leave most of the power in the tube generator. Obviously, a matchmaker is required. We find one in the *ratio* of primary to secondary turns of the output transformer.

A large number of primary turns is matched to the tube, and a much smaller number of secondary turns to the speaker. The primary coil will have a *high* voltage and *low* current flow; if a high voltage is needed to push a small amount of current through the circuit, the opposition, or *impedance,* must be considerable. The turns ratio of the transformer provides a proportionately *low* voltage with *high* current flow in the secondary (page 38); as a lower voltage pushes a much larger current through the secondary, the opposition, or *impedance,* must be relatively small.

Suppose we have an output stage with a power tube whose recommended load impedance is 5,000 ohms, and we want to use an eight-ohm speaker. The *impedance ratio* would be $5,000/8 = 625$. But to wind the transformer, we need to know how many *turns* to use for the primary and the secondary. We learn this by dividing 5,000 by 8 and finding the square root of the result. $\sqrt{5,000/8} = 25$. Thus, if we need 3,000 turns for the primary, our *turns ratio* of 25 tells us that 120 turns are correct for the secondary $(3,000/25 = 120)$.

The actual *load* on the output stage is really the speaker's voice coil, which is busily-occupied converting its electrical energy to mechanical energy; in other words, in vibrating the paper cone. Only a small per cent of the electrical power emerges as acoustical power. Different types of speakers vary greatly in efficiency. Efficiency, however, is not a measure of quality.

The secondaries of most of today's output transformers are tapped for use with speakers of different impedances. This transformer would have an 86-turn secondary tap for the four-ohm speaker. $(\sqrt{5,000/4} = 35. \ 3,000/35 = 86)$. A tapped secondary is not quite as efficient in matching as one with all its turns in use; and a transformer designed for a single speaker impedance is preferred.

Actually, when the manufacturer says his speaker is a four- or an eight- or a 16-ohm speaker, it's only an approximation. The ohmage is roughly half the direct current resistance and half the inductive reactance—at some selected frequency, usually around 600 cycles. Therefore, the impedance will drop considerably at the low end of the frequency spectrum, and increase at the high end. Then there is the speaker's resonant frequency to contend with. This is the frequency at which the voice coil and its cone diaphragm tend to vibrate freely like a struck tuning fork. The impedance rises steeply at this frequency, because our loudspeaker *motor* has turned into a *generator*. Vibrating on its own in the strong magnetic field, the voice coil generates a voltage that feeds back into the amplifier.

Generally speaking, the loudspeaker's resonant frequency should be placed as far down as possible. This is more easily done in the case of the *woofer,* which is designed to reproduce the bass notes only. Many such speakers have resonant frequencies as low as 20 cycles.

The transformer should have enough primary turns to prevent impedance from falling off too much at the low frequencies. Remember that the impedance of a coil increases with frequency. The primary also has a resonant frequency. The transformer, therefore, should be designed so this resonant frequency is above the audio spectrum.

The impedance load in the plate circuit of triodes may be from two to three times greater than the plate impedance. This avoids distortion by helping to keep the signal on the most linear portion of the characteristic curve.

The optimum working load for the pentode in a power amplifier ranges from 1/4th to 1/10th of the plate resistance. The load line for either triode or pentode is drawn for a resistive load. The actual line, which can be pictured on the screen of an oscilloscope (the same as TV's picture tube) under operating conditions, is elliptical, because the impedance is constantly changing with the complex of changing frequencies. A poorly designed output transformer can be a prime source of distortion.

For the push-pull operation, now universally used with hi-fi, the optimum working load is different. With triodes, you can use a load impedance closer to the impedance of the two tubes together. As we have noted, push-pull cancels out most of the even harmonic distortion. With pentodes, the load impedance is usually about double the amount for a single-end amplifier.

The more efficient pentodes can deliver several times the power of a pair of triodes. The price for this greater efficiency is distortion, though when the even harmonic distortion is controlled, *negative feedback* can be added for cancelling out most of the remaining odd harmonic distortion.

Negative feedback is to hi-fi what the pneumatic tire is to transportation. The principle was developed by an early Marconi radio operator, Stuart Ballantine (1897-1944). The invention is described in U.S. Patent No. 18,835, granted in 1923. The same patent includes automatic volume control, which is a form of negative feedback. In the same year, Louis Alan Hazeltine, American physicist, adapted the idea to the broadcast receiver called the *neutrodyne*.

The only available amplifier in those days was the triode. At radio frequencies, the capacitance between plate and grid of a triode offers little reactance, and the positive voltage feedback from the plate to the grid circuit is often high enough to make the tube begin to oscillate. Hazeltine engineered a circuit that fed back just enough *negative* voltage to cancel the positive feedback. This quieted the whistles caused by the heterodyning of the signal frequency with those generated in the receiver. During World War II, the principle was widely used in radar, guidance systems, computers, and other electronic devices. It has been less than a decade since the audiophiles first seized upon negative feedback as a cure for non-linearity in the audio frequency amplifier. It is sometimes called inverse feedback, or degeneration.

At first, one might think that feeding an output voltage back to the input "in reverse," to cancel distortion, would be futile because the tube would nullify it again. But a tube doesn't *amplify* the distortion, it *generates* it. The distortion is not present in the input. Imagine a pure sine wave applied to the grid of an amplifier. The output contains a third harmonic distortion—the two half cycles may be affected like the sine wave pictured in Fig. 50 (page 158). So we take a fraction of this output voltage, and bring it back to the grid, 180 degrees out of phase with the pure sine wave input (Fig. 10B). Because this inverse feedback is directly opposed in phase, it distorts the sine wave in the opposite way in which the tube distorts it. Through this method, close to 100% linearity can be achieved, and any noise and hum reduced.

Negative feedback has its price too. It always reduces the *gain* of an amplifier in the same ratio that it reduces distortion. If distortion is reduced by a factor of 10, say from 5% to .5%, then you need roughly 10 times more input voltage than before. This may mean an extra stage of amplification.

The loss of gain results because we can't feed back the distortion alone. Much of the signal voltage must also be returned to the grid.

Inverse feedback is also not without its dangers. The signal, in passing through the amplifier, changes phase because of coupling capacitors and stray capacitances. As a result, feedback may not be exactly 180 degrees out of phase with the signal but something less. It may even be enough less to amount to *positive* feedback, creating an opposite effect to the one desired. Instability, even oscillation, can follow and even though the oscillation frequency may be above the audible, it can still cause the music to sound harsh and unreal. To counteract this, a small capacitor, in parallel with a resistor, is placed in the feedback loop (Fig. 54).

Negative feedback is always most beneficial to an amplifier

that is already a fairly good amplifier without it. This is why some perfectionists still prefer triodes to pentodes. The Williamson amplifier uses triodes in push-pull all the way, both for voltage and power amplification.

As shown in Fig. 54, the feedback loop usually is taken off from the secondary of the output transformer in order to include the distortion from this device's non-linearity.

Well-engineered feedback also helps the frequency range. Amplification normally falls off some at both ends of the audio spectrum. Feedback is proportional to output. The higher output of the middle frequencies means that they suffer a proportionately greater loss of feedback, leaving the two end frequencies on higher ground than before.

The many virtues of inverse feedback become vices when the amplifier is overloaded. When power output is increased beyond the rated working limit, distortion jumps up like a startled cottontail. This is especially true of pentodes. Since most hi-fi amplifiers today use pentodes in push-pull, with a large dose of negative feedback, this should be a warning against turning the volume up too high.

Before we end this brief excursion into the mysteries of the output stage, we should describe *ultra-linear*. In 1937, A. D. Blumlein, an engineer with E'ectric and Music Industries (EMI, Ltd., London), patented a circuit that combined the high-power efficiency of the pentode with many of the low-distortion virtues of the triode operation. In the November 15, 1951, issue of *Audio Engineering*, Hafler and Keroes reintroduced it as the *ultra-linear amplifier*.

What is this magic circuit? There are two methods for using pentodes in the output stage. Connected as triodes, the plates and screen grids are tied together. Connected as pentodes, the screen grids have their own separate connection with the power supply, with resistors to reduce the voltage somewhat, and decoupling capacitors to keep the voltage as steady as possible. The ultra-linear hook-up differs from both.

The screen grids get their voltages from taps on the output transformer. (See Fig. 54.) Each screen grid tap is taken off at a point that gives it about 43 per cent of the turns of its half of the primary. The ultra-linear operation is obviously *in between* that of the triode and pentode hook-ups.

Working as a triode, the screen grid voltage swings along with the plate voltage; as a pentode, the screen grid voltage is kept as constant as possible while the plate voltage does all

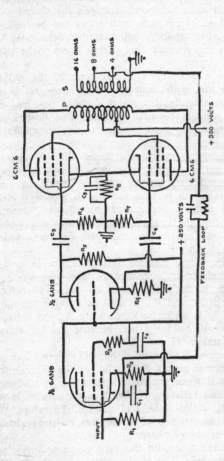

FIG. 54. Diagram of the main (power) amplifier for high fidelity. The pentode voltage amplifier (6AN8) and the triode phase inverter (6AN8) are in the same envelope, a multipurpose tube. The coupling between them is direct (see Fig. 44). The two voltages for the push-pull output tubes, which must be 180 degrees out of phase, are taken from across R_4 and R_8. Coupling to the grids is through C_3 and C_4. The screen grids of the output pentodes are connected to taps on the transformer (Williamson circuit). R_5 and C_5 provide the cathode bias for these tubes, which are 6CM6s. 6L6s or 5881s could also be used. R_6 and R_7 are grid leak resistors. The capacitor across the resistor in the feedback loop is very small; it compensates for the phase shift of the signal that results from passing through C_3 and C_4.

the swinging. Ultra-linear splits the difference: the screen grid swings, but its voltage is somewhat lower than the plate voltage. The internal resistance of the tube is greatly reduced.

Operational results also are a compromise. With a slightly higher plate voltage supply, ultra-linears put out almost as much power as pentodes. And with deviation from optimum load, such as often occurs on loud passages, the circuit generates less distortion than it does with either the pentode or triode operation. Most quality output transformers sold today have taps on the primary that permit them to be used in the ultra-linear circuit.

Next, let's have a quick look at the *input* of the output stage. We realize that with push-pull tubes, one grid is always going negative while the other is going positive. To obtain opposing voltages requires a circuit that inverts the phase of one of them—as good a reason as any for calling it a *phase inverter*.

There are probably as many as a dozen different circuits that will do the job by employing either one or two tubes. If two tubes are used, they may be enclosed in the same envelope. (This is the multipurpose tube; each group of electrodes includes a cathode for a separate electron stream.) The diagram of Fig. 54 (page 169) illustrates a popular, single-tube hook-up, called the *split-load inverter*, which also includes a *cathode follower*.

R_4, R_5 and the tube's internal resistance are in *series*. We know one of electricity's elementary laws—current flow in a series circuit is the same in all its parts. Therefore, the current through R_4 and R_5, as well as through the tube, will be the same. So if we make R_4 equal to R_5, the voltages across them also will be the same, because voltage equals current times resistance. ($E = IR$.)

The voltage across R_4 also will be 180 degrees out of phase with the voltage across R_5. Earlier in the chapter, we learned why it is that an amplifier inverts the grid voltage across its plate load resistor. The voltage of a cathode resistor, such as R_4, follows the grid voltage. Therefore, the voltages taken from the top of these two resistors always will be 180 degrees out of phase.

It should be noted that with this type of phase inversion, there is no voltage gain. The explanation is simple. In Fig. 27 (page 103), capacitors are connected across the resistors in the cathode circuits. We know that the capacitor's job is to *smooth out* the audio frequency, leaving a steady d-c voltage. However, in the circuit of Fig. 51, we are *using* the audio frequency, and without the capacitor, this frequency across the cathode resistor is fed back to the grid 180 degrees out of phase. Actually, a cathode resistor that has not been bypassed offers another means of obtaining negative feedback or degeneration. Therefore, any voltage gain from the tube is nullified.

Of course, if we want our hi-fi cake frosted with stereo, we need a *pair* of amplifiers like the one described above. This is the ideal, though you can combine the two amplifiers by using separate input tubes and output transformers. The phase inverter and output tubes will handle both signals (sum and difference signals) from stereophonic record or tape.

It would be a relief to many, including the author, if we could junk the whole subject of amplification at this point and discuss loudspeakers and their enclosures. But the high road to fidelity has long demanded a second amplifier, called the pre-amplifier. This one may be a separate unit, or it may be mounted on the same chassis as the power amplifier.

The amplifier needs nearly one volt of input to operate with good volume and without distortion. Crystal and ceramic pickups generate this much voltage, or more. But most hi-fiers demand the pickup that generates its voltage through electromagnetism, and these low-output devices (one to 70 millivolts) require a pre-amplifier ahead of the amplifier.

Electromagnetic pickups fall into two broad classifications —moving coil and moving iron. The former is similar to the moving coil dynamic microphone; a stylus instead of diaphragm furnishes the motive power for the coil. The coil vibrates in the magnetic field of a permanent magnet. The ESL, Fairchild, and Grado are of this type.

In the moving iron pickup, the coils, or coil, are stationary. Manufacturers use various methods to generate a voltage in the coils by a magnetized iron armature, to which the stylus is attached. The vibrating armature may disturb the magnetic flux between the two poles of a magnet on which the coils are wound, or the armature may be a magnet, which vibrates inside a coil. There are other methods as well. Moving iron pickups include the Pickering, General Electric, Audak, and the Shure.

Crystal and ceramic cartridges utilize the piezoelectric effect. Since 1880, scientists have known that when pressed (*piezo* means "to press"), twisted, or deformed in some way, certain crystals will generate a voltage. A Rochelle salt crystal (sodium potassium tartrate), grown from a supersaturated solution, has long been used as the generating element in both phonograph cartridges and microphones. In 1946, the Sonotone Company developed a ceramic (barium titanate), also a piezoelectric, which has certain advantages over Rochelle salt when used in a cartridge.

Improved models of both these cartridges deliver from 20 to 1,000 times the voltage of the electromagnetic types, and their frequency response is also adequate. Since no coil is necessary, they are immune to hum pickup. Then why not use them and jettison the pre-amplifier? Most commercial set manufacturers do use them. The audiophiles advance two arguments against them: they are not always immune from *needle talk,* and they may add some "coloration" of their own to the music. (Needle talk consists of sounds produced by vibrations originating in some part of the cartridge.)

Piezoelectric cartridges have other disadvantages. They are usually cheaper than the electromagnetics, and without preamplification, they require fewer knobs. But the hi-fi manufacturers aren't in business to go out of business. And —so the story goes—if the man who assembles his own rig settles for anything less than the electromagnetic cartridge —well, he's hardly the kind of a man you'd want your sister to marry.

Ceramic piezoelectrics include the Astatic, CBS-Hytron, Electro-Voice, Erie Resistor, Sonotone, and the Webster Electric. The Duotone and Ronette are Rochelle salt crystals. Some of the new two-element piezoelectrics for stereo have done much better in the quality market than the old mono type. Others have been built to a price for the mass market.

The preamplifier has two (or, at most, four) stages of R-C-coupled voltage amplification, sometimes with feedback. The feedback may be both negative and positive—negative for erasing distortion and positive for increasing gain. In addition, the preamp usually incorporates numerous control circuits, which make it the nerve center of the high-fidelity complex. (The *basic* diagram of a preamplifier is shown in Fig. 55.)

One of the preamp's most essential circuits is the one that provides *equalization.* To understand the need for equalization, we must make a quick visit to the recording studio, where high fidelity begins.

Programs are usually recorded first on magnetic tape so

that they can be edited before they are transferred to the record. Transfer to the record is accomplished through a transducer called the recording *head*. The head changes the audio frequencies to electrical vibrations which, in turn, operate the cutter. The cutter engraves its vibrations in a spiral groove in the acetate. It is a *constant-velocity* device, which means that the cutter's side-to-side motion always maintains the same speed. To do this, it must move further for the low frequencies than the high ones.

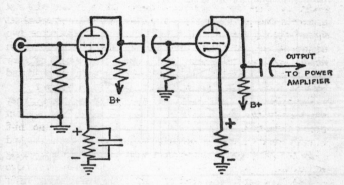

FIG. 55. Basic two-stage preamplifier. We do not see the equalization, gain, tone, and loudness controls, which are usually connected between amplifiers and ground.

The long-playing 33⅓-RPM records measure 325 grooves to the inch, each groove no wider than a human hair. The cutter stylus cannot move as far from side to side for the low frequencies as it would normally tend to do; there simply isn't enough room. To prevent the stylus from crossing the boundary line to adjacent grooves, a filter circuit reduces the energy of all frequencies below 500 cycles.

The result of this change from constant velocity to constant amplitude below 500 cycles is greatly reduced output from the record. The amount of reduction is around three to five db per octave; each time the frequency is halved, the energy is only half or less its previous strength.

Above 2,000 cycles, we are faced with another recording problem. For these frequencies, the cutter stylus need move but a relatively short distance from side to side. And beginning around 5,000 cycles, the surface noise, which has been mounting all the time, is almost equal in volume to the music.

The solution at this end of the scale is the opposite to the one at the other end: increase the energy to the cutter at around three db per octave above 2,000 cycles. This operation is called treble pre-emphasis.

Thus the record we carry home from the record shop with such Himalayan expectations has been "rigged"; the bass has been reduced in level, the treble boosted. The preamplifier must reverse the results of this rigging by boosting the bass and cutting (de-emphasizing) the treble. This is what is meant by *equalization* (Fig. 56).

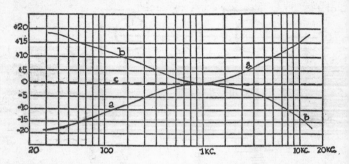

Fig. 56. RIAA Recording Characteristic, adopted in 1953. In recording, the bass is cut and the treble boosted, as indicated by line *a*. An equalizing circuit in the preamplifier boosts the bass and cuts the treble, as shown by line *b*. Line *c* is the result.

Actually our equalization circuit doesn't really *boost* the bass, which could only be done by means of a resonant circuit, which is hardly practicable in this instance. The feat is accomplished in the same way that we might "raise" an underwater island in a lake. We raise the island by draining off enough water to leave its surface above the lake's level; the equalization circuit "drains off" enough high-frequency energy to leave the bass on an even keel with the rest of the frequency spectrum.

We accomplish this through filtering. The filter circuit of Fig. 57, which is usually connected between the two amplifiers and the ground, absorbs a certain percentage of the current of all the frequencies, but the amount removed increases as the frequency rises. In other words, the highs pass through the filter (to the ground) more easily than the lows, so we call it a high-pass filter.

As we know, an effective high-pass filter is either a capacitor in the line, an inductance across the line, or both in combination. Our equalization high-pass filter uses capacitors and *resistors*. A resistor offers equal opposition to all frequencies; a capacitor's opposition rises as the frequency drops. For example, capacitive reactance (X_C) of a .01-μfd capacitor to 10,000 cycles is close to 1,600 ohms; at 5,000 cycles, it is 3,200 ohms; at 1,000 cycles, 16,000 ohms; and at 100 cycles, 160,000 ohms. And we shouldn't forget that when we add this reactance to a resistance, as in the series connection of Fig. 57A, the addition is algebraic.

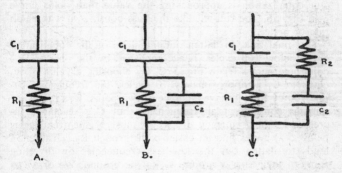

FIG. 57. The equalizing circuit for a preamplifier is illustrated by (C). The first two stages, (A) and (B), are shown merely to help clarify the explanation.

Fig. 57A gives us most of our equalization below 500 cycles. If the capacitor's reactance is equal to the resistor's resistance at 500 cycles, the signal voltage will divide equally across them. But below 500 cycles, the increasing reactance of the capacitor doubles each time the frequency is halved (each octave), as stated above. This means an increase of around three db per octave.

On the other hand, the current-to-ground opposition to the frequencies above 500 cycles is increasingly smaller. These frequencies are "drained off," leaving the lower frequencies in the amplifier. The *level* of the frequencies above 500 cycles hasn't been changed much; but the reduction of all these frequencies has, in effect, progressively boosted everything below 500 cycles, as drainage raises the island above the level of the lake.

But a bass boost that leaves the treble above 500 cycles almost *flat* in comparison is not sufficient. The entire bass must be boosted still more, and at the same time the treble must be cut (de-emphasized), as indicated by line b in Fig. 56. Fig. 57B shows a second, smaller capacitor connected across R_1. Then the frequencies above 500, which pass through C_1 without much opposition, can choose between R_1 and C_2.

C_2 offers increasingly less opposition as the frequency rises; therefore, more and more of the current of the higher frequencies passes through it to the ground. The result is a treble *roll-off* to the high end of the audio spectrum. At the same time, the complete bass is left "higher and drier" than before. Our equalization is approaching the value that yields curve b in Fig. 56. The roll-off, like the bass boost, is at a three-db rate or better.

Our final task, to flatten out the slope of the equalization curve below 100 cycles, is accomplished by the resistor, R_2, connected across C_1 in Fig. 57C. By choosing a value of R_2 equal to C_1 at approximately 70 cycles, the frequencies below 100 cycles will have a choice of two paths. R_2's resistance remains constant at all frequencies, but C_1's reactance increases as the frequency drops. So most of the frequencies below 500 cycles find an easier path through R_2. This action tends to flatten out the low-bass frequencies by diverting some of their energy around C_1 to the ground. We arrive at our final equalization curve, b of Fig. 57, which complements the recording curve a. The addition of these frequencies gives us the dotted line c, which is close to flat for the whole audio spectrum.

The total loss in amplification through equalization is around 20 db. Many other circuits, using different combinations of resistors and capacitors, may be used for equalization in the preamplifier.

If all recordings began the bass roll-off and treble pre-emphasis at the same points, called cross-over points, we would need only a single, fixed compensating circuit. But standardization began only as recently as late 1955. Since that date, all of our records have used the recording curve of the RIAA (Recording Industry Association of America). The bass roll-off starts at 500 cycles, the treble pre-emphasis at 2,000 cycles, although the cross-over points are actually rounded off as illustrated in Fig. 56.

Other compensating curves are necessary for older recordings and foreign labels. The compensating switch might include two more filter circuits—a scratch filter (high pass) for old and worn records and tape hiss, and a rumble filter (low pass) to correct hum and rumble. These two filters are likewise unkind to the musical frequencies. The scratch filter takes a slice out of the highs, the rumble filter bites into the low lows. However, the rumble filter can operate effectively down around 30 cycles, where you'll scarcely miss a loss of frequencies, even if the speaker is capable of reproducing them.

We have been discussing equalization for an electromagnetic cartridge. Like all electromagnetic generators, the voltage produced by this type of pickup is directly proportional to the velocity. However, if we use a piezoelectric pickup, the voltage generated is directly proportional to the degree to which the piezoelectric element is bent or twisted, that is, to the distance the stylus moves from side to side.

As the recording stylus still moves further from side to side for the bass than for the treble, we are confronted with a different equalization problem: we must *cut* the bass and *emphasize* the treble. This can be done by simply connecting a resistor of from 1.5 to two megohms across the cartridge. Unlike the electromagnetic variety, this cartridge is a high-impedance device and requires high-impedance matching. It also has some capacitance, which, together with the big resistor, provides us with a low-pass filter to ground for both bass cut and treble boost.

The preamplifier includes other necessary controls. Next in line is a selector switch, which may be combined with the switch for the equalizer circuits. The selector makes available the proper impedances for such sources as the electromagnetic cartridge, the piezoelectric cartridge, tape, FM radio tuner, AM radio tuner, TV, and microphone. Each input has its own impedance matching and volume problem.

Any preamplifier has an input level control, a volume control similar to a radio's gain control. In addition, a *loudness* control is always included. This one consists of a number of resistors and capacitors, which may be even more complex than the equalization circuit. Its function is to adjust automatically the volume to the human ear *for a different loudness level.* As I pointed out earlier, the ear favors the middle frequencies over the bass and the highs—except at a loudness level of 90 db, or higher. Even then, the frequencies above 5,000 are still down, but this isn't as important as the bass. Some loudness controls compensate only for our ears' weakness to the bass.

Most preamplifiers also include separate bass and treble controls. Room size or room acoustics may dictate the use of these controls. You must be particularly careful when you advance the bass control because too much power in these frequencies can easily overload the amplifier. Distortion also results if the speaker can't reproduce much bass.

Since the 1958 debut of the stereophonic record, for many of us, stereo has become one of life's necessities, ranking close to multiple vitamins and Elizabeth Taylor. The store-bought phonograph with no more than a single speaker is almost as old fashioned as the kind you used to crank. Stereo units selling for under $100 come equipped with as many as four saucer-size speakers and the salesman's promise of such miracles as "flat from 30 to 15,000."

I have heard no complaints from the buyers of this merchandise. They don't seem to miss the bass or resent the muddiness. In fact, since the evolution of the phonograph, most people have been happy with whatever music they could obtain from the latest model they could afford. Only the perfectionists are unhappy. The disc stereo boom was first promoted by the manufacturers of components for home assemblers. They require, among other things, a *pair* of the preamps described above, mounted on a single chassis.

With a separate preamplifier for each stereophonic channel, you could retain two full sets of monophonic controls. No doubt, many frustrated engineers would relish both the flexibility and the complication this provides, but most owners prefer a modicum of simplification. This may be had by *ganging* the controls for the two channels, as we do with tuning capacitors, so that they turn in unison. A single knob simultaneously adjusts both channels for equalization, volume and loudness control, source selector, bass, treble, et cetera. Each manufacturer has his own ideas as to which functions should be ganged and which operated separately.

Most of the newer stereo preamps use what is called the clutch-type control. This consists of two concentric knobs, the largest one against the panel, which can be adjusted independently. You can also turn them in unison by pressing in the outer knob. Thus, you can adjust each channel separately, and then, after you have achieved a proper balance, you can operate both together. This arrangement can be effective with bass and treble controls when the dual speakers aren't identical, or, if identical, they have slightly different characteristics.

The stereo preamp also requires some extra controls and switching of its own, the most important being balance control. To obtain good balance between the two channels, a pair of volume controls is ganged in such a way that, as one channel's gain is increased, the other's gain is decreased. Overall loudness doesn't change; it shifts from one speaker to the other. Both speakers should have equal volume.

If you are stuck with an old-fashioned mono disc, such as Beethoven's *Ninth* directed by Toscanini, can you play it on your beautiful, new *haute monde* stereophonic rig? Yes indeedy, and although no stereo effect will be present, two speakers always sound better than one. (By the same token, three sound better than two, et cetera.) But for minimum distortion and noise, the dual outputs from the cartridge should be fed into the preamp in parallel, and this requires another switch. While it is still best to use a mono cartridge with the mono disc, this cartridge never should be used with stereo discs. The stereo channel requires up-and-down as well as side-to-side motion, which most mono cartridges can't supply, and this lack of *vertical compliance* wears out the record.

Simplification has advanced to the point where today's manufacturers mount the preamp on the same chassis with the main amplifier. Most manufacturers are now including both AM and FM stereo tuners with this combination. The music lover can choose from more than 50 different makes of these *integrated* stereo preamp-amplifiers now on the market, priced from $54.95 to $289.50. Power output *per channel* ranges from 13 watts to 100 watts. Power output should be chosen with regard for the efficiency of the speakers. Speaker efficiency is not always the twin of quality, as we shall see in the next chapter.

12

THE WHY OF HI-FI: THE LOUDSPEAKER

THE SPEAKER, or system of speakers, for either monophonic or stereophonic usually is regarded as the most important stage on the road to hi-fi heaven. It is the final stage, of course, and it usually leaves a deeper mark on the reproduction than any other. It can ruin the output of a well-nigh perfect amplifier, and it often can make an inadequate amplifier sound pretty good. Here is where the audiophile gets the most for his money in pleasing re-creation of sound. He can economize to better advantage on the cartridge, arm, turntable, preamp, and amplifier than he can on the transducer. To be more precise, a medium-priced turntable, arm, and cartridge, and an economical preamp-amplifier, will serve naturalness more effectively than an indifferent speaker and enclosure. However, let's not push this theory too far, especially in the matter of the phono cartridge.

The moving-coil speaker described in earlier chapters is still one of the pillars of high fidelity (Fig. 58). The *electrostatic tweeter* has earned a place for itself in the field, but the *full-range* electrostatic is still a novelty and very expensive. In the electrostatic loudspeaker, the diaphragm vibrates in response to *voltage* changes. In the moving coil device, the coil's varying magnetic field reacts with the fixed field of the permanent magnet to move it in and out. As the voice coil's current, and its consequent field, varies with the frequencies of the original sound waves, these waves are reproduced in the motion of the voice coil and its attached cone diaphragm.

The vibrating diaphragm not only must reproduce all of the varieties of sound from an orchestra's percussion, woodwind, brass, and string instruments; it also must handle them all mixed up together with a convincing degree of naturalness.

Is this moving-coil dynamic speaker some kind of miracle instrument? How closely it approaches the miracle category depends upon good engineering (which is largely synonomous with high price) and the type and quality of its enclosure.

As I am always pointing out, the speaker is a motor. Following Volta's chemical battery, and the development of electromagnets, experimenters began seeking ways for utilizing the *force* in electromagnetism. Joseph Henry built a reciprocating motor. Its iron bar armature, pivoted in the center, rocked back and forth over a pair of electromagnets, one under each end of the bar. Current through the two coils attracted one end of the bar armature, and repulsed the other. As the bar rocked to one side, it reversed the direction of the current through the coils, changing their polarity, and causing the bar to come down on the other side. This action would again reverse the direction of current flow with the same results.

Today's telephone receiver has a diaphragm of some magnetic material, essentially the same as Henry's soft iron bar, that vibrates in response to the varying current through a pair of coils. In principle, then, it is also a *moving iron* device.

The first radio loudspeaker was simply a large earphone with a horn added. It not only rattled when a loud signal caused the diaphragm to touch the pole pieces; it distorted badly, and delivered practically no bass below 300 cycles and no treble above 2500 cycles.

Improved designs followed, including one that used a balanced armature, something like Henry's motor, between the poles of a permanent magnet. The signal went to a coil wound on the armature. A small cone diaphragm, attached to the armature's outer end, would reproduce bass notes down to 120 cycles. But speakers were still in the iron age, and when a *moving-coil* speaker appeared in 1926, with its startling reproduction of both bass and treble, a new day in sound reproduction had arrived.

Around 1860, William Thomson (Lord Kelvin) used a moving-coil galvanometer, with a tiny mirror attached to the coil, to read the feeble signals from the first Atlantic cable. The first moving-coil *speaker* ever built may be seen today

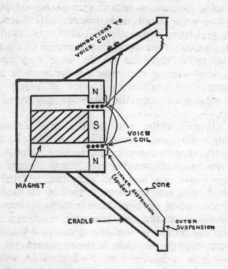

FIG. 58. Cross-sectional view of "old reliable," the dynamic (moving coil) speaker, which today may be called the PM, for permanent magnet speaker. The center pole magnet, now made of Alnico VI, of high intensity, is surrounded by a mild steel of high permeability (low opposition to the magnetic force). Two magnets may also be mounted at the outer edges of the assembly. The voice coil is made of very fine wire, and consists of many more turns than are indicated by the dots in the drawing.

in the Science Museum at South Kensington, near London. Its patent, No. 9,712, was filed on April 27, 1898, by Sir Oliver Lodge. Lodge intended it to be used on battleships as a "loud hailer" in a public address system. The headphones then in use were the moving iron type borrowed from the telephone. Hi-fi moving coil (dynamic) headphones for mono or stereo, now available on the market, are ideal for quality reproduction. Stereo headphones provide a high degree of realism.

In the early twenties, Chester W. Rice and Edward W. Kellog went to work on a moving-coil speaker at the G.E. laboratories. Because only weak permanent magnet material was then available, they were forced to use an electromagnet. The coil of an electromagnet of sufficient size uses from 20 to 40 watts, too big a load for a battery set. So after building a moving-coil speaker with six-inch cone, Rice and Kellog provided it with its own amplifier and power supply. The amplifier had the unprecedented output of a whole watt; the power supply that replaced the old A and B batteries consisted of a rectifier and filter for use with 110 volts a-c (Fig. 28; page 107). Their *Radiola Model 104,* for plugging in to the home power supply, came on the market in 1926, priced at $200.

A few years later, all radios and phonographs, except portables, were playing with moving coil speakers, and all except the portables had a plug-in power supply. Bell Laboratories also developed this new, vastly-superior dynamic speaker. Power from the house current solved the problem of the speaker's electromagnet. But the hum from the coil had to be neutralized, the unit was bulky and expensive, and a good permanent magnet replacement was desirable.

The first permanent magnets available for speakers were too weak for good efficiency. One, which had six per cent of chrome added to the steel, weighed five pounds. An equal strength cobalt steel magnet that followed halved this weight. A few years later, a greatly improved alloy of aluminum, nickel, and cobalt, called *alnico* made its appearance. Today, an alnico magnet with the same flux strength as the old five-lb. chrome magnet weighs only a couple of ounces; but the alnico magnet in a quality speaker may still weigh five pounds or more.

The advantages of the high-flux density in these speakers are many. For one thing, the mechanical force is proportional to the voice coil current and flux density. For example, an increase from 5,000 to 17,000 gauss can raise the sensitivity nine db. This is equivalent to increasing the power output of the amplifier eight times, which means five watts at 17,000 gauss will do the work of 40 watts at 5,000 gauss. The efficiency increase extends into the highs, raising the output in the 2,000-10,000-cycle region something like 10 db. On the lows, the type of enclosure must be considered along with the *damping factor* to be discussed in a moment.

High-flux density gives the varying signal from the amplifier a much firmer grip on its voice coil. An improved *transient response* is one reason for the crisper and more brilliant reproduction. A transient is a sound current that begins and stops abruptly, such as from a bang on drum or cymbal. The start of a transient can produce a sound peak called an *overshoot*. This is not as objectionable as what follows its stop—an independently vibrating cone. The term for it is *ringing* or *hangover* (Fig. 59). A flux density of at least 10,000 gauss is necessary to effectively "snub" low-frequency hangover. Hangover caused by resonances of portions of the cone or its mountings are mostly in the high-frequency range and must be taken care of by careful design.

Transients are also a problem in amplifiers. You can check them by applying a square wave to the amplifier input, and noting its shape on the screen of an oscilloscope connected to the output. It should emerge substantially square.

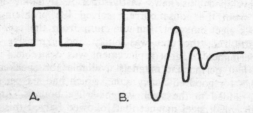

Fig. 59. Transient distortion that uses square wave. The steep drop-off in the intensity of the sound causes the voice coil to continue to vibrate as it does at B. This *hangover* sound mixes with the notes that follow with ill effect. The movement of the voice coil also may lag the steep rise of the leading edge (left side) of the sound wave. A blow on the snare drum creates a wave similar to the square wave. If it sounds blurred, the transient response is poor. High flux density is the best guarantee of good transient response.

Cone resonance creates a problem very similar to transient response. A musical tone that is close to the natural, fundamental frequency of the cone also can start it vibrating on its own. This produces an unnatural, thudding bass, as welcome as a cry in the night. Such a vibrating voice coil becomes a *generator* instead of a *motor*. The generated voltage sends back a current through the transformer to the output tubes.

Disconnect one of the leads to the speaker and push in gently on the cone with your finger. It moves easily against the inner and outer suspension. Tap it, and it will vibrate at its fundamental frequency. Next, connect a piece of wire across the terminals to *short* the voice coil. Now, try giving the cone a quick push, and it will resist. The movement of the voice coil's turns of wire across the lines of force of the magnet's powerful magnetic field generates a voltage in the coil. The current from this voltage, moving through the voice coil and negligible resistance of the short, is accompanied by a magnetic field, which opposes the field of the permanent magnet. The *faster* you push the cone, the higher the voltage generated; the larger the current flow, the *harder* you must push. Also, the higher the flux density, the harder you must push. If, in place of the direct short across the voice coil, this circuit contained considerable resistance, there would be less current flow and less opposition to your push.

Low impedance triodes provide the necessary low resistance, which is called a *high damping factor*. But pentodes with sufficient inverse feedback can be equally efficient. A high damping factor also helps control hangover from low-frequency transients.

To sum up the case for high-flux density, its greater efficiency requires less power input from the amplifier, and, consequently, less risk of distortion on loud passages; its high-frequency performance is an improvement; its transient response is better, resulting in a crisper, cleaner sound; the cone resonance is more effectively controlled. It looks as if there's no substitute for flux—unless it's a miracle material for the *cone* (and its suspension elements).

Before we take up the cone, I'd like to add one final word about the voice coil—namely, distortion. As in an amplifier, all kinds of non-linearities can originate in the speaker. As we know, non-linearity and distortion are closer than gin and tonic.

One of the main causes of voice coil non-linearity is a non-uniform magnetic field. One end of the coil moves outside the field on long excursions. Since the low notes alone cause the long excursions, the resulting distortion is confined to the bass. It trims off some of the top and bottom of the wave (Fig. 60). This amplitude, or harmonic distortion, can be cured by a voice coil that is either so much longer (or so much shorter) than the fixed magnetic field of the permanent magnet that even for long excursions the magnetic relationship between them doesn't change. However, the cure sacrifices efficiency, because *some* of the over-long coil's magnetic field is wasted and so is *some* of the permanent magnet's field in the case of the too-short coil.

The early speaker cones were simply oversize paper drinking cups. However, the improvement over the old horn speaker was so dramatic that it took a little time for the

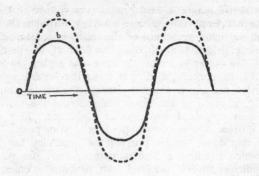

Fig. 60. The manner in which a loudspeaker whose voice coil moves out beyond the maximum magnetic intensity in the gap can trim off the top and bottom of a sine wave. The ingoing wave *a* emerges as *b*.

cone's faults to surface. For one thing, the cone didn't reproduce a satisfactory range of frequencies, even though the range was wider than the horn's. A small cone, say three or four inches in diameter, handles the high frequencies best, a large cone is best for the lows. As we increase the cone size to the point where it has good low-frequency response, the highs are left behind . . . or almost, as we shall see shortly.

A large cone operates like a piston and pushes more effectively against a room's air pressure than a small cone does; its larger surface area gathers in more of the air. Increasing the *mass* of the cone also helps the bass at the expense of the treble.

At 100 cycles, a 12-inch cone may move as much as one-fifth of an inch, peak to peak, on good loud volume (one acoustic watt). On the other hand, for the smaller surface of an eight-inch cone to deliver this concert hall volume at 100 cycles, it would have to move three times as far. As we have seen, non-linearity can be a serious problem with such wide excursions, and it is next to impossible to obtain good, high volume bass response from an eight- or ten-inch speaker without serious distortion.

Cone mass (weight) is chosen for the frequency range it is designed to handle. The wood pulp used varies widely, particularly in regard to the length of the fibers. Some manufacturers mix kapok, or wool, with the pulp to achieve the relationship of mass with the stiffness or flexibility of its suspension. It is this relationship that determines the cone's fundamental frequency. An increase in mass lowers this frequency; an increase in suspension stiffness raises it. The formula is $f = 1/2\pi \sqrt{\dfrac{stiffness}{mass}}$; where f stands for frequency, stiffness (or suspension) is in dynes per centimeter, and mass is in grams. I shall describe the suspension system in a moment.

We still can enjoy a good speaker, with a fairly wide frequency range, if we don't ask for too much volume. The high frequencies move the cone a very short distance, and most of their radiation comes from the cone's apex next to the voice coil. For this reason, the apex can be made of a very stiff material, using the heavier, soft-textured pulp for the large, outer section of the cone only. Such a coupling confines the highs to the inside, while the bass frequencies push and pull the entire cone over greater distances. Another method for separating the two frequencies is to use a small, separate cone for the highs. The high-frequency cone may be attached to the apex of the larger cone, or it may have its own voice coil, in which case it is called a *coaxial* speaker.

Cheaper speakers often increase the frequency range by using a corrugated cone, which "breaks up" for the higher frequencies. However, this makes for a very uneven response in the treble range.

I once had a car that would start to "talk" to me at exactly 27 miles per hour; this speed appealed to the fundamental resonance of some hidden part. We all know that a sound frequency that is precisely equal to the fundamental of a water glass can shatter it. Now consider the many electrical frequencies of varying amplitudes that activate a loudspeaker. One of the designer's most difficult problems is to prevent the natural resonances of the speaker's numerous parts from distorting these frequencies and causing hangover.

As we know, the fundamental resonance of the complete cone is determined by the relationship between its mass and the stiffness of its suspension. The cone requires both an outer and an inner suspension. The inner one also must hold the voice coil exactly in the center of its narrow gap. (See Fig. 58; page 182.) Early models were shaped like a *spider*, and were so named. Most manufacturers now use a corrugated disc, made of fabric or paper. Its center is glued to the voice coil, its outside edge to the metal frame or *basket,* as it is called. The disc type also protects the gap from dust.

The natural resonance of the spider may be more of a problem than its effect upon cone resonance. It must be rigid enough to prevent the voice coil from moving off center to scrape against the magnet's poles; on the other hand, too much rigidity adds more than the law allows to the cone's natural frequency.

If the low frequencies push the spider too far, they can damage it. Replacement is seldom practicable.

The outer suspension is also a challenge to the designer. The mass-manufactured corrugated cones are usually attached directly to the outer ring of the basket. But some kind of suspension material between cone and ring produces better results. Rice and Kellog used rubber in 1926. Different kinds and grades of leather, and many kinds of cloth, have been used in the past. Today foam rubber, impregnated cloth, as well as other materials are used, each manufacturer having his own ideas as to which is best for low cone resonance, linear excursion, and smooth treble response.

As in the amplifier, amplitude distortion also can create *intermodulation distortion* in the speaker. A frequency that is amplitude-distorted, usually down in the bass, modulates a higher frequency to produce other frequencies (beating together, heterodyning; see page 152) that are harsh to the ear.

The cone also tends to vibrate at a number of other frequencies higher than its fundamental. These frequencies, whose origin is rather complex, show up in a frequency response curve (Fig. 61). Some frequencies are accented, but, more frequently, a narrow band of frequencies is largely cancelled. (The cancellations aren't often noticeable.) The term for this phenomenon is cone breakup, as I have mentioned before. The partial remedy is a soft, long-fibered pulp for cone material.

The problems confronting the design engineer, who sets out to build a speaker with a wide-frequency range, low distortion, and sufficient volume, should be clear enough by now. Though it is relatively expensive, one way out of the dilemma long has been to divide up the frequency spectrum between two or three separate speakers. It is much simpler to design a low-distortion speaker for a limited frequency range. The lows are handled by a *woofer,* the highs by a *tweeter,* and, if a third speaker is used, a *squawker* takes care of the mid-range. However, a very satisfying response still

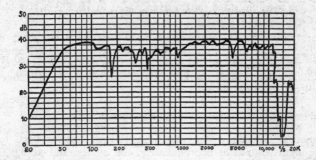

Fig. 61. Diagram of a typical response curve for a speaker mounted in an enclosure.

can be obtained from a single, quality speaker of compromise size.

First, let's look at frequency range. The high-flying hi-fier speaks reverently of frequencies down to 30 cycles, as if they were clusters of emeralds. Listen to that bass, man! Most of the time, our *aficianado* isn't even getting the low notes he thinks he is. Very few instruments have bottom notes under 60 cycles, and with the exception of the organ and the symphony orchestra's double-bassoon, double bass, and bass tuba, seldom produce them. The cello's lowest fundamental is 70 cycles, the trombone's 85. Probably 99 per cent of the music on today's records have no more than a trace of these deep-down notes. And when they do appear, their second and third harmonics usually represent them quite adequately. And the bass under 70 cycles is almost always distorted in the speaker.

The human ear can be happy with a limited frequency range, provided it's *balanced*. From a center frequency of around 800 cycles, the range should extend equally on both sides. Imbalance on the treble side sounds strident, on the bass side boomy. For example, moderately-priced "high-fidelity" packaged phonographs, or radio-phonographs, may have a restricted range of from 100 to 5,000 cycles. If the usual distortion were absent, this range would be very pleasing. Of course, a range between 80 to 8,000 without serious distortion would be more pleasing, and with a single quality speaker this range is attainable from an assembled hi-fi rig in the medium price range. Even 65-70 to 10,000 is possible at not too great cost.

Clean, crisp reproduction of a limited-frequency range is much preferred to a wider range with more distortion. The ultimate true high-fidelity range, 30 to 20,000 or even 20 to 30,000, has a price tag that frightens most of us. Much of this money goes into the bass enclosure or horn.

Adequate dispersion throughout the room can be a problem with single-speaker hi-fi. The bass radiates nicely in all directions, at a 180-degree angle, to fill the room. But as frequency rises, directivity sets in. A 10-inch speaker starts to *beam* its output at approximately 400 cycles; and at 5,000 cycles, you would have to sit directly in front to hear this frequency well. Of course, the highs bounce around so well that they can be dispersed by hard surfaces; the speaker's output can be beamed at a wall or ceiling for improved dispersion. A corner location also helps.

Single-speaker advocates are more numerous in Great Britain and on the continent than in this country. With a single speaker, the room can't be too large or too noisy, because volume must be kept at a moderate level—no two-block amplifier. An eight- or ten-inch speaker is preferred, and in most instances, its resonant frequency should be below 70 cycles. (A speaker's output falls off rapidly below resonance.)

The owner can't be bass-happy either; he must sacrifice some of the *complete* naturalness of cymbals, drums, mariacas, and the "way-out" harmonics from Louis Armstrong's trumpet. However, the fidelity of this moderately-priced hi-fi system should be emotionally satisfying to 99.44 out of 100 listeners. A *pair* of these speakers with stereo amplifiers offers increased naturalness by adding depth and directional effect.

Now we come to the business of mounting the speaker in a good enclosure. Anyone who has ever been wakened in the night by a cat meowing in the basement, two floors below, well knows that sound waves can't be stopped by anything short of solid concrete walls several feet thick. If you stand in a canyon between two deep, rocky walls and shout, you realize that there is rubber in sound waves too! They bounce back and forth between reflecting surfaces.

When you bang a drum or toot a horn in a room, some of the acoustic energy *passes through* the walls, which also *absorb* some of it; the walls also *reflect* part of the energy, and any portion can cause sympathetic vibration of objects in the room.

If you aim your speaker directly at a good sound reflector, such as glass, plywood, or cinderblock, the reflected wave will contain considerable energy. Sound waves combine as electrical waves do, by adding or subtracting on the basis of phase. The combination of reflected wave with direct wave creates a *standing wave,* which can be weaker or stronger than the direct wave.

The standing wave's strength is determined by the speaker's distance from the wall and the length of the wave. (To obtain wavelength from frequency, divide the speed of sound by the frequency; thus, $1130/40 = 28$ feet for a 40-cycle wave.) The standing wave will consist of alternating peaks of intensity (loops) and depressions (nodes) where the sound is inaudible. The loops and nodes account for uneven listening levels throughout a room.

The subject of *room resonances* is a fascinating one. The distance between two parallel walls, and between floor and ceiling, determine these resonances. For example, a room 11 feet wide has one fundamental frequency of 50 cycles. Fifty cycles is a wavelength of 22 feet, and a half wavelength, 11 feet, can start the reflections bouncing back and forth to create standing waves. There will also be harmonics of 100 cycles, 150 cycles, et cetera. The other two walls, the floor and ceiling also have their resonances. *Combinations* of distances between walls, and between walls and ceiling and floor, produce still other resonances.

The larger the room, the lower the frequency of the room resonances. In the average-sized room, on the order of 12 x 15 feet, most of the energy in the resonances is confined to the long wavelengths—those of frequencies under 100 cycles. The worst possible room dimensions would be a perfect cube, in which all the resonances combine and reinforce one another. The ideal size for a room with an 8½-foot ceiling is said to be 13½ by 21 feet.

Drapes, carpets, sound-proofed ceilings, and overstuffed furniture contribute to the damping of room resonances; they absorb the higher frequencies better than the lows. But there should be enough reverberation left for the required "liveness." In mathematical terms, the *reverberation time* should be a little under one second at 500 cycles. A reverberation of longer duration begins to sound like an echo.

This brief survey of room acoustics makes it clear why no music system sounds the same in all rooms. Experimentation in placing the speaker—and adding to, or removing, absorbent material, such as drapes and carpets—is usually necessary.

Our discussion of room acoustics brings us to the subject of loudspeaker enclosures, most of which are nothing more than tiny "rooms," into which we pump a high volume of sound. The two main problems with these miniature rooms are the "room" resonances and vibrating panels.

Why is the speaker enclosure needed? Well, no gadget has been devised that will vibrate in one direction only. The *back* movement of the speaker cone radiates almost as much sound energy as the *front* movement. Because of this, the low frequencies, which disperse so widely and build up so slowly, present a problem. Their back and front waves have time to meet, out of phase, and largely cancel each other. Unless we manage to keep them apart, the bass all but disappears.

The simplest way to do this is to mount the speaker in a sheet of plywood, called a *baffle*. Theoretically, the speaker would have to be mounted in a seven-foot-square board to prevent cancellation down to 81 cycles. The distance of seven feet between the speaker's front and back is equal to one half-wavelength of the 81-cycle frequency. (1130/81 = 14) (14/2 = 7). Half a wave separation is enough to prevent cancellation.

Practically, a smaller board would be almost as efficient. As the speaker usually sits on the floor, there can be no return from its underside, which reduces the amount of cancellation by one-fourth. The distance around the corners is greater than seven feet, which also helps. Then much of the back radiation, including bass as well as treble, will be reflected from the wall behind the speaker.

The old commercial consoles, which were with us for so many years, used an open-back cabinet, which is a folded version of the flat baffle. Although the bass was false and boomy, as well as uneven, those old three-watt sets put out a lot of sound, much of it reflected from the wall behind.

An improvement over this simple baffle, whether the board alone or a board folded back into an open-back cabinet, is the speaker mounted in a wall between two rooms. The wall or door of a closet may also be used. A quality speaker with a powerful magnet, properly mounted in a wall, can reward us with an even, fairly-wide frequency response of satisfying quality. As none (or very little) of the rear radiated frequencies are returned, this is called an *infinite baffle*.

We can arrange a substitute infinite baffle by using a box

for the other room. Because of its relatively small size, the walls of this box must be lined with an absorbent material to soak up the back frequencies as completely as possible. The highs are quite effectively absorbed; the lows are not absorbed as easily. The size of the box also affects its operation. The smaller it is, the more its air pressure stiffens the cone. And this increased stiffness raises the cone's natural frequency, thus limiting the bass. The dynamic speaker puts out very little energy below its fundamental frequency.

Cone-damping is actually turned to advantage, however, in a small enclosed book-shelf speaker invented by Edgar M. Villchur. This air-tight box utilizes the "spring of the air," as it was called by Thomas Boyle (1627–91), author of the first treatise on electricity. Here at last is something for hi-fi that is truly linear: the atmosphere's pneumatic spring.

I mentioned above the distortion caused by a speaker's mechanical suspension (spider and rim) with large voice coil excursions for bass notes. When pushed or pulled too far, the mechanical suspension can't follow the magnetic force quite far enough and some of the wave is "shaved off." Usually the same amount is removed from the wave's top and bottom, causing odd harmonic distortion. But if more is removed from one of the half cycles than the other, the asymmetrical shape means even harmonic distortion. It's similar to clipping by an amplifier when it's operated beyond its characteristic curve.

Villchur's "acoustic suspension" substitutes linear air for the non-linear mechanical suspension. In his words, the speaker has purposely been made "defective," been "crippled," with an extremely loose outer suspension that leaves the cone as floppy as the seat of grandpa's pants. This drops the initial cone resonance to 10-15 cycles, which is raised to 40 cycles by the stiffening caused by added air pressure. Good bass is produced down to this frequency, though without too much volume or dynamic range. Its efficiency is also very low, caused largely by the use of a voice coil longer than the magnetic gap to reduce distortion. Thus, the amplifier used with this convenient little speaker must develop from 30 to 50 watts of power. It is also expensive to buy: from Acoustic Research (AR), KLH, or Heath.

With the conventional speaker, the box should be as large as practicable to prevent an increase in cone resonance. A nine-cubic-foot box doesn't stiffen the cone appreciably, but resonances vibrating panels are still a problem.

Enclosure resonances produce much higher frequencies than listening room resonances because of the relatively shorter distances between the sides, top, and bottom of the box. A cube would be the worst choice. It is helpful to use a box with as many panels of different size as possible in order to spread the standing waves over a wider spectrum. The corner type of Fig. 62A is a good example.

The smaller panels of the corner box also help to minimize vibrations. Padding is added to help the absorption, but the vibration still persists. The spurious frequencies add something to the music, and at the same time cause sharp dips at various frequencies. They can make a violin sound like a violoncello, or even a cello.

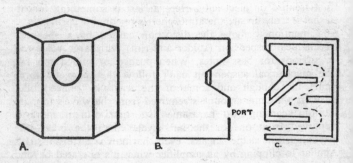

FIG. 62. Three types of loudspeaker enclosures: (A) a corner enclosure; (B) a side view of a bass reflex cabinet, showing the port; (C) the side view of a labyrinth enclosure.

Some experts recommend a double wall lined with sand to minimize vibration. A "house of bricks" is better yet. The British authority Mr. G. A. Briggs is noted for his corner brick enclosures. Brick walls don't vibrate like panels, and the music comes out relatively sweet and clean. A nine-cubic-foot model, with the same shape as the panel job of Fig. 62A, has a 12-inch woofer, with the mid-range speaker and tweeter, mounted on top, looking up at the ceiling. This model has a *port*—an opening beneath the woofer—which introduces us to a different type of enclosure than the completely-enclosed box (Fig. 62B).

The wall or box-mounted speakers that I have discussed may be termed *direct radiator* types. A second opening, mentioned above, permits the controlled use of some of the energy from the *back* motion of the cone to extend and reinforce the bass response. The first type of this enclosure was the *bass reflex,* which began to find favor among distortion-conscious audiophiles in the early forties. Its relative cheapness, simple construction, and small size, account for its wide popularity. The initial patent on this ported cabinet probably belonged to A. L. Thuras, of Bell Laboratories, in 1930. The Jensen company gave the invention its name in 1936.

The bass reflex method of backing up the bass utilizes the Helmholtz (1821-1894) discovery that a cavity with a small vent, or port, will resonate at a particular frequency. Helmholtz worked out the mathematical equations for this resonator, in which air moves in and out of an opening; the elasticity is provided by the air compressed in the cavity. Blow across the mouth of a jug, and it responds with its resonant frequency, which is determined both by the area of the jug and the size of the opening. The shape of the cavity is not important.

If a single, wide-range speaker is used in a bass reflex, it had best be an eight-incher. Speakers of 10- and 12-inch diameters sound better with a tweeter for the highs. The box size is determined by the size of the speaker. The smaller the box, the closer the air coupling inside between the cone's rear motion and the air that moves in and out of the port. And the smaller the box, the higher the cone's natural frequency. With a 12-inch speaker, the enclosure required to keep down the cone frequency rise should not be less than seven cubic feet; 10 to 12 cubic feet is preferable.

The box resonance is determined by its volume and the size of the opening. Its volume is equivalent to the capacitance (C) in an electrical circuit; the size of its opening is equivalent to electrical inductance (L). You should choose a box resonance equal to the speaker's natural free air resonance.

The added stiffness of the cone, caused by the air confined in the box, raises the speaker's natural frequency. Fig. 63 shows a speaker's free air resonant peak. Mounted in the bass reflex cabinet this peak is reduced to the two much lower peaks, one on each side. The lower frequency peak extends the bass. Both peaks are levelled off somewhat by a quality speaker with high flux density and an amplifier with high damping factor (page 184). The speaker cuts off rapidly below the lowest peak because of the mechanical resistance of the suspension system. Now let's see just how the bass reflex operates to extend the bass.

The air moves in and out of the box through the port at its resonant frequency; and it would seem that the air would be forced out by the cone moving in and vice versa. As a result, the two sound waves would be 180 degrees out of phase, causing cancellation. But because of the inertia of the air in the box, which is influenced by the size of the port (L), the outward motion of the air is delayed by close to one half cycle, bringing the outputs into phase and adding. However, this only happens with the low frequencies, on each side of the speaker's open-air natural frequency.

Higher, faster-changing frequencies *do* emerge out of phase with the speaker's directly-radiated energy. For this reason, you should pad the enclosure well to *absorb* these higher frequencies as completely as possible. The box frequencies below the cone's free air resonance also emerge somewhat out of phase with the cone, which accounts for the lower impedance peak. Despite this drawback, there is enough in-phase air to add to the bass.

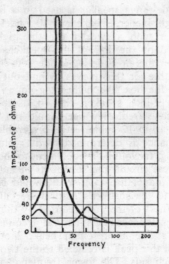

Fig. 63. (A) We see the normal impedance of a speaker, taken in free air, showing the steep rise at the resonant frequency of the cone. It is usually not as high as this. (B) Illustrates the twin peaks that result from mounting the speaker in a bass reflex enclosure; note how the bass is extended. The peaks are suppressed by a high damping factor. Impedance also rises considerably at frequencies above the 200 cycles indicated.

Vibrating panels and cavity frequencies still contribute to distortion and coloration in the bass reflex. Listening tests indicate that the smaller panels in the corner enclosure, with no two faces parallel, provide the most pleasing reproduction.

You can also extend the bass by funneling the cone's back radiation through a labyrinth. The principle of an *acoustic labyrinth* is simple. A pipe or conduit will serve, though it is much easier to build shelves inside a wooden cabinet, over and under which the sound must wend its way. (See Fig. 62C; page 194.) The distance from back of the speaker to the opening is one quarter of a wavelength at some low frequency close to the resonance of the speaker when mounted. For 60 cycles, this is 4.7 feet. By this dodge, the backwave energy in the bass region is delayed long enough to emerge in phase with the front radiation.

An extra dividend accrues, as it does in the bass reflex, from the fact that equal volume in the bass region is attained with reduced excursion of the voice coil. The difficulties of construction make the acoustic labyrinth less popular than the bass reflex. It may also house a single, full range eight-inch speaker.

All the enclosures, from simple board baffle, or open-back box, to labyrinth, are meant to preserve and extend the bass notes. The treble is exposed to all sorts of buffets between vibrating panels and standing waves that cause distortion and add a "boxiness" to the response. Isn't there some ideal way to reproduce the sound of music in the home? Let's take it from the top again, and see if we can come up with something better.

The relationship between the vibrating cone and the air of the room is one of the hurdles in understanding high fidelity. Hi-fi is both electronics and acoustics and, different as they are, they still share many concepts as in the case of the bass reflex. We think of the *coupling* of the cone with the air as a problem of *impedance match*. For maximum transfer of energy the cone size and frequency should work into an atmosphere impedance that is closely equal to it. However, it's only the lows that give us trouble in this respect.

The larger the cone, the better the match with the air for the bass frequencies. Since the speaker size is limited, you can create the effect of a larger cone by mounting a number of speakers close together, the more the merrier; the output at *all frequencies* adds, thereby boosting the bass to a pleasing level.

If, at the same time, the treble becomes too prominent, you can control it easily by a simple high-pass filter to ground. Small, relatively-inexpensive speakers, such as five-inchers, may be used. Because the individual excursion of each voice coil is short, distortion is minimized, and the speakers may be mounted in a fully-enclosed, well-padded box no more than six inches deep. You should employ an even number of units, from four to 16, so that each row can be connected in series, and the rows then connected in parallel. This series-parallel connection permits one to match the speakers' total impedance with the amplifier. But given a single speaker, impedance match with the air can be greatly improved by the use of a horn.

The speaker's diaphragm meets the air in the horn's narrow neck, where the air, being restricted, has something "solid" to work against. Thus the diaphragm's power is not wasted in futile motion against low-pressure air. At the same time, the mouth of the horn becomes, in effect, the area of a cone. The horn is really an acoustic amplifier.

Then why aren't *all* hi-fi sets equipped with horns? Many of them are—for the middle and high frequencies. Most tweeters have horns, both for an improved cone-to-air impedance match and as a diffusion booster. But only a small horn is required for the short, quick excursions of the cone at these frequencies. For the much larger, longer excursions of the lows the horn must be a big one—in order to effectively *load* the cone. Shape and size are also important considerations. Among the infinite variety of possible shapes the exponential shape offers the best compromise results between distortion and low-frequency response. However, these are all technical points, which are "off limits" for this book. For a horn to operate efficiently with a frequency as low as 40 cycles the area of the mouth would have to be close to 64 square feet. If the mouth were square, this would mean eight feet by eight, and the length would have to measure around 12 feet.

Obviously, few houses, not to mention apartments, could be fitted conveniently with a horn of this size. But the home-building audiophile who chases after higher and higher fidelity can include one of the big horns in his plans. He should build the horn first, followed by a large room growing out of its mouth. The room's dimensions should be similar in size to the one mentioned earlier as close to ideal.

The final touches would include drapes, and police dog

to fend off irate neighbors. Fifty watts of power, fed into one of these highly efficient bass horns should, in contrast to de Forest's two-block amplifier, qualify the rig as a twin-city amplifier in most areas. But the owner would still be frustrated in this hi-fi age. No room for stereo!

The large horn speakers in use today are what are called corner horns. The speaker is mounted in a cabinet which is designed to provide the first section of a folded horn, something like the acoustic labyrinth. The horn's final section utilizes the corner walls of the room. The flare is hardly ideal but results are very satisfactory. When a horn is attached to the rear of a speaker in this way, the speaker is said to be *back-loaded*.

The first of these corner horns was the Voight, named after its British inventor, introduced about 30 years ago. One of today's best-known corner horns, which is licensed to several manufacturers, is the Klipsch (Fig. 64), described originally in the *Journal of the Acoustical Society of America* for October, 1941. Mid-range and high frequency horns are mounted above.

When the frequency range is divided among two or three speakers, each speaker must have its own separate channel.

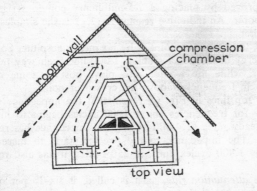

FIG. 64. Top view of the Klipsch corner horn, for extended bass reproduction, first described in 1941. The sound waves pass through a multi-cell construction (similar to the labyrinth of Fig. 62C) for a short distance before utilizing the floor and corner walls for the flareout. The air mass in the horn, against which the front of the cone works, necessitates surrounding the cone's rear with a compression chamber to balance the pressure on both sides.

This calls for a dividing network, more often called a cross-over network since you always choose a definite frequency for "crossing over" from one range frequency to another.

Such a network uses the high- and low-pass filters that we first mentioned in Chapter 4. The tweeter channel would be a high-pass filter, the middle speaker channel a band-pass filter, and the bass channel a low-pass filter.

Because capacitive reactance (X_C) declines as the frequency rises, a single capacitor, in series with the tweeter, makes the simplest high-pass filter. Above a certain frequency, determined by the size of the capacitor, the reactance is so low that it almost amounts to a short circuit.

The capacitor size is determined by the cross frequency you choose and the speaker impedances, which should be equal. For a 1,000-cycle crossover frequency, with 16-ohm speakers, 12 μfds is about right; with eight-ohm speakers, 24 μfds is okay. These values equalize the input impedance.

At the crossover frequency, the capacitor protects the small tweeter from the low frequencies, which, in moving the voice coil too far, could easily wreck the inner suspension; at the same time, the woofer remains open to the highs as well as the lows, taking about half their power. This situation may be improved by placing a 2.5-millihenry coil in series with the woofer. An inductive reactance (X_L) is the simplest low-pass filter.

Still, the crossover point is by no means absolute. Some of the energy in frequencies below 1,000 find their way into the tweeter, while the woofer also plays host to some of the energy in frequencies above the crossover point.

The top lines in Fig. 65 reveal that the impedance to the woofer for frequencies above 1000 cycles begins to increase at a steady rate. At 4,000 cycles, the impedance has reached 15 db. The impedance to the tweeter starts to increase at roughly 1,800 cycles until, at 250 cycles, it has also reached 15 db.

This *attenuation rate*, as it is called, is six-db per octave. We remember that a three-db change, which is just barely noticeable, results from doubling (or halving) the power. An octave represents the doubling (or halving) of the frequency. In this instance, each time the frequency is doubled on the way up, or halved on the way down, the attenuation has increased by six db. And a six-db increase means that the power is only a fourth of what it was before.

We also learned in our initial discussion of filter circuits

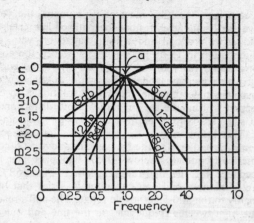

Fig. 65. Diagram of the roll-off for three different crossover networks with a crossover frequency of 1,000 cycles (point a). The current to the low-frequency speaker rolls off to the left; the current to the high-frequency speaker goes to the right. The 6-db roll-off usually is considered insufficient, the 18-db roll-off too great. A 12-db roll-off, obtained with the circuit of Fig. 66A, is the most popular.

that a capacitor, or coil, connected in parallel with the circuit, acts in reverse to one connected in series. This enables us to obtain a much *sharper cutoff* by doubling the attenuation rate to 12-db per octave. We simply place a second capacitor in parallel with the woofer and a second coil in parallel with the tweeter. If this sounds complicated, you can clear up your doubts by following the highs and lows through the

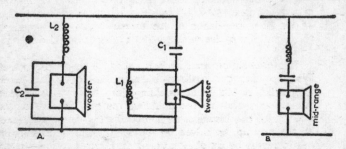

Fig. 66. (A) Crossover network for a 12-db roll-off. (B) Network for the mid-range speaker.

diagram of Fig. 66A. Like most of us humans, the voltages that move the frequencies follow the path of least resistance.

By adding a third coil and a third capacitor, you can increase attenuation to 18-db per octave. However, too complex a network is bad for transient response, and some engineers say that it also increases phase distortion to such an extent that it becomes objectionable. Your best solution is to choose speakers that will handle the *overlap* without distortion.

Three speakers require a third channel for the mid-range speaker. This means *two* crossover points. For instance, the tweeter may be assigned to the frequencies above 5,000 cycles, the woofer to frequencies below 500. A bandpass filter lowers the impedance for the mid-range speaker (Fig. 66B). This filter can combine a coil and capacitor in series that "tunes" very broadly to the 500-5,000-cycle range. The coil's reactance keeps the high tweeter frequencies in the line and out of the speaker; the capacitor's reactance does the same for the low-woofer frequencies.

13

THE WHY OF HI-FI: STEREO

SINCE mid-1958, the magic word that fills the audio *aficionado* with new desire and fresh zeal, while emptying his pocket-book, has been *stereophonic*. For a time, he was lolling on Cloud Nine, enjoying the "utmost" in hi-fi equipment, with some tiny compromises here and there in deference to price tags, when there suddenly appeared on the horizon, no bigger than a man's hand, the 45/45 stereo disc. The disc required a double order of almost everything—pre-amplifier; main amplifier; speaker, or set of speakers; and pick-up cartridge.

His old turntable or changer still carried the twin-channel record solo. However, rumble had become more of a problem with the stereo pickup, which must respond to vertical as well as lateral motion, and for that reason, he was faced with the possibility of having to replace the turntable.

Rumble is one of a trio of "distortions" originating in an imperfect turntable and no turntable is 100% perfect. It consists of low-frequency vibrations from the motor picked up by the cartridge. Rumble can be heard alone with no programming. Its companion wreckers are *wow* and *flutter,* both caused by a wavering motor speed, slowly for wow, creating a perceptible change in the pitch of a tone, more rapidly for flutter.

What is stereo's secret? With mono, the music that comes pouring out of a single "hole" can become flat and uninteresting, even tiresome, to the sensitive listener. In a word, monophony can become monotony. On the other hand, in the concert hall, our attention is diverted continuously from one section of the orchestra to another as strings, bass, percussion, woodwinds, or a combination of these, relieves the full orchestra of the burden of the theme.

Monophony, it would seem, is like watching a parade through a knot-hole; or a view of the beach that is restricted to the area where a single wave attacks the shore with cosmic patience. Through honest trickery, stereophony would accommodate the home listener, by placing him in the 15th row center of the concert hall.

The principle of stereophonic sound first was demonstrated in 1881, during the Paris Exposition. On opposite sides of the stage at the Paris Opera stood a carbon microphone. From each microphone, a telephone line carried the sound to a headphone in a room that was blocks away. By wearing the headphones as a *pair,* the listener's ears became, in effect, the two spaced microphones on the opera stage.

What the carbon mike and those early headphones did to the fundamentals, not to mention the harmonics, wasn't too important at the time. At any rate, the *binaural* (two ears) effect was present. It was demonstrated to the public again in 1932, with much improved equipment, at the American Academy of Music in Philadelphia. A tailor's dummy named Oscar was fitted with microphones for ears and placed on the auditorium stage. The output from the mikes was amplified and fed to pairs of earphones worn by the visitors. One year later, Oscar was a featured performer at the Chicago Exposition. The eight-inch space between the microphones, corresponding closely to the space between headphones, worked much better than the wider separation used in Paris.

The general public was introduced to stereophonic sound via 35-millimeter motion picture film. Walt Disney's *Fantasia* pioneered the multi-audio-channel reproduction now taken for granted in movie houses. Cinerama obtains its startling, realistic sound effects through seven channels; Cinemascope uses five; the standard for wide-screen pictures is three. This is termed the *curtain of sound* principle—as many microphones as possible, with an equal number of loudspeakers at the other end for something approaching naturalistic reproduction in hall or theatre.

Since speakers are not yet available in the handy six-pack at the supermarket, some means of simplification had to be sought for home stereo. Fortunately, it was discovered that a single *pair* of speakers can provide a satisfactory measure of the stereo effect in a small room.

Following World War II, we discovered that the Germans

were ahead of us in recording sound on magnetic tape. Their
Magnetophon recorder served as a model for American machines made by Ampex, and their plastic tape was adopted
and improved by Minnesota Mining and Manufacturing, and
the Armour Research Foundation. It was the latter's Marvin
Camras who patented the use of red oxide as the magnetic
element on the tape.

Soon the slower-moving (7½ inches per second) red oxide
tape, with good frequency response up to 15,000 cycles, made
binaural tape, used with headphones, available to the audiophile—provided he could afford a tape recorder with twin
heads. The same red oxide, with a special binder, provided
by Minnesota Mining and Manufacturing, to cement it to
the tape, is used in the more recent video tape recorders from
Ampex.

At the recording studio, two microphones, eight inches
apart and separated by a board, simulate the human ear.
When this system, which is effective with headphones, is used
with speakers, the listener is confined to a small area, with
only some three feet to spare on each side, in order to profit
from a stereophonic effect. A number of recording techniques
have been developed since for use with speakers. Before I
describe any of them, we should examine the way in which
our ears locate the origin of sounds.

We may use one or more of at least four different methods
for locating a sound's source—(1) time difference, (2) intensity difference, (3) phase difference, and (4) ratio. There
is a slight difference between each ear in a sound's arrival
time, though we might have to turn our heads slightly to determine this. The width of the human skull, from six to eight
inches, can cause a maximum difference of half a millisecond
(the speed of sound is 1130 feet per second). This is more
time than the brain needs. In fact, the difference can be as
little as six microseconds. The uncanny cooperation between
our computing brain cells and our ears can break down in a
heavy fog, as is evident from the numerous collisions that
take place in London.

The second clue to the origin of a sound is found in the
relative intensities of the sounds that reach our ears. For example, if our left ear "looks" towards the source, that ear will
hear a somewhat louder sound than will the right ear. The
difference is greatest at the high frequencies, though there is
no agreement among the experts as to where the treble starts:
250 cycles, 500, 700? The difference in loudness is con-

sidered to be more revealing than the time it takes for a sound to reach each ear.

Our ears also can note a difference in phase. A cycle of sound, consisting of a compression and a rarefaction, is located at a different point in the cycle for each ear (provided, of course, that the wavelength is at least as long as the distance between our ears, which would be the wavelength of a frequency of some 1500 cycles). According to tests, the lag, or lead, between our ears can be as little as 100 microseconds.

The ratio mentioned above is the one between direct and reverberated sound. Stereo makes good use of this ratio to provide a sense of spaciousness. At a concert, the ratio of reverberated sound that bounces off walls, ceiling, and floor, to the direct sound, may be as high as 90 per cent. Figuratively speaking, the listener swims in a sea of sound. Some of this sense of involvement can be achieved at home by regulating the loudness of the direct sound relative to the reverberation. Of course, too much reverberation can muddy up the waters, or even degenerate into an echo. The latest hi-fi gimmick is a feedback circuit in the amplifier that adds to the reverberation.

Whichever method informs the listener—time, intensity, phase, ratio, or combinations of these—his head, like a radar antenna, is never still. Its involuntary movements are constant, if slight. Tests at Bell Laboratories reveal that these movements are necessary to enable him to *externalize* a sound source.

The stereo recordist must never forget the human ear. He may use the so-called *classic* method of recording, based upon the curtain of sound principle, with variations. Two microphones are placed in front of the sound source, anywhere from six to 30 feet apart, depending upon the program material. For a jazz combo, the recordist would select the six-foot placement; for a symphony orchestra in a large hall, he might use the 30-foot separation. With this system, both time and intensity differences provide the stereo effect.

As regards time, the sound from the left of the stage would reach the right microphone (R) milliseconds after it has passed by the left microphone (L). To be effective, this difference should be at least three milliseconds, representing only about three feet. Anything above 50 milliseconds supposedly can provide a sensation of two distinct sounds rather than an impression of direction. The difference in the intensity of the sound from the two speakers also contributes to the listener's sensation of lateral dimension, though to a much smaller degree.

This simple method suffers because the recordist has no assurance that his reproducing speakers will be spaced the same distance apart as his microphones. And except for small groups, the left-right differentiation may be too pronounced for a realistic sense of participation. Consequently, he must consider a number of other problems, including not only the program source, but the size and acoustic character of the hall or studio, the type of microphone, and other puzzlers.

The closer the mikes are to the sound source, for example, the greater the ratio of direct sound to reverberation, which contributes to the effect of spaciousness. When you move them back, you achieve more spaciousness, but at the cost of directional effect. In his home, the audiophile often must move closer or further away from his speakers to obtain the most realistic reproduction from a recording.

If too great a distance between microphones is required for this time-intensity-reverberation technique, a third, or center, microphone may be helpful. By recording all three channels, the center mike's signal may be added later to one or another of the L and R signals, at the discretion of the editor.

A number of other very ingenious recording techniques have been devised for creating the stereo effect. One of the most interesting of these is the German M-S system. M-S stands for the German "front on" and "sideways." We get around this by calling it Mid-Side recording.

Two capacitor microphones are mounted in a single housing (Fig. 67). Because they are so close together, the full sound reaches them at practically the same instant. This eliminates time or phase clues, leaving only relative intensities to indicate directional effect.

The mikes are placed front and center before the sound source. The mid-mike (M) has a cardoid (heart-shaped) sensitivity pattern, which means that it picks up everything from the center (C) as well as from the right (R) and left (L), though not with exactly equal volume. The diaphragm of the side mike (S) is mounted at right angles to the mid-mike; its sensitivity pattern is like a horizontal figure eight. The S mike's front pickup is very small, increasing as the angle swings around toward each side. Thus it is essentially a "side-picker-upper."

Of course, the S microphone can provide only a single

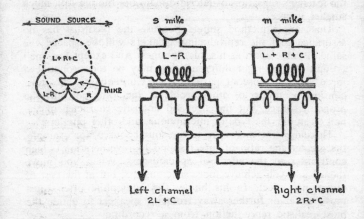

FIG. 67. The mid-side recording system for stereo uses two capacitor mikes in a single housing. The mikes have the directional properties indicated on the left. The stereophonic effect is attained through intensity differences. By combining the outputs of the two mikes both in phase and out of phase, as indicated, the left and right channels suitable for stereo emerge.

channel. Sound reaching it from the two sides fight it out on the basis of phase. "Out of phase" means that the sound from one side pushes, while the sound from the other side pulls. In effect, we obtain a different signal. The output from the left (L) is subtracted from the output from the right (R), or vice versa, according to the direction of current flow. Let's see what this can mean.

The M mike's output goes to the primary winding of a transformer, as does the S mike's output. Each transformer has two secondary windings (Fig. 67). This makes it possible to combine the outputs from the two mikes either in or out of phase. Such a circuit is called a *matrix*.

The M (center) mike with its wide pickup delivers the full program, $L + R + C$. The S mike delivers, in phase, $L - R$. Adding the two we get:

$$\begin{array}{r} L + R + C \\ L - R \\ \hline 2L \quad\quad + C \end{array}$$

To obtain the other channel we reverse the connections of

the S mike's secondary coil. This changes $L - R$ to $R - L$. Adding again, we get:

$$L + R + C$$
$$-L + R$$

$$\overline{}$$

$$2R + C$$

So we end up with an L channel $(2L + C)$ and an R channel $(2R + C)$, both of which carry information from the center. And we accomplish it with a pair of capacitor mikes in a single housing placed in front of the sound source. For reproduction, instead of side-by-side speakers that we use, the German M-S system uses one speaker directly facing the listening area (for the M mike), and one speaker turned sideways (for the S mike).

Although stereo on tape was with us before disc stereo, the latter was actually developed first. Britain's A. D. Blumlein, who has been called the father of modern stereo, was recording two channels in a single groove in 1930. Early records used the vertical, called the "hill-and-dale" method of recording. The audio frequencies moved the cutting head up and down, the loudest signals cutting deepest into the groove. The vertical was replaced by the lateral (side-to-side) method, which is less subject to distortion. In 1931, Blumlein patented a stereo disc that used the vertical for one channel and the lateral for the other. Some of his early discs are still around, and sound excellent on a modern reproducer.

The high cost of tape stereo in the beginning prevented its mass acceptance. Making copies from a master tape is a much slower and more expensive process than pressing records. However, such is the competition from disc stereo that the price of tape has dropped, and the new *two-way* stereo tapes, which double the playing time, may be cheaper, note for note, than stereo records. The new tape cartridges are also convenient to use. With reasonable care, a tape will survive many more replays without damage than a record, and its quality is generally superior. But a good tape recorder, for playback, costs hundreds of dollars; and the more familiar records, with their far wider choice of program material, still comprise the bulk of stereo sales.

By the time the public was ready to accept stereophonic records, the Blumlein patents had run out. The Westrex 45/45 method, which Blumlein had also envisaged, was finally chosen from among several systems. This method employs both vertical and lateral movements with a unique feature of its own. The two channels are recorded, one on each of the two walls of a V-shaped groove. Each groove wall is at an angle of 45 degrees to the vertical, which is why it's called the 45/45 system. The walls are therefore 90 degrees apart.

How does the pickup stylus manage to extract the necessary information from both walls of the groove without moving in two directions at once? If the single cutting stylus can *record* the two channels in a single groove, a single playback stylus should be able to *reproduce* them, which returns us to the recording studio . . .

The recording head converts the electrical forces of the two channels into mechanical forces. At the same time, the magnitudes and directions of the two forces are combined. The net resulting force moves the cutter either vertically or laterally, or in some compromise direction, to carve the frequencies in the two walls of the V-shaped groove.

If the sound from the L channel is the louder its force will be greater, and the cutter will record it, for the most part, in the L wall of the groove. If the sound from the R channel predominates, it will proceed mostly to the R wall.

Suppose the studio microphones pick up sine waves of equal magnitude but opposite in phase. The cutting head resolves the two out-of-phase forces by moving back and forth (laterally), leaving an equal amount of signal in each wall. The stylus of the pickup cartridge is mechanically coupled to two transducer units mounted at right angles to each other. As the stylus moves laterally, one transducer generates an opposite voltage. On the other hand, if the two sine wave signals are in phase, the cutting head moves vertically; and the pickup stylus, following the up and down grooves, generates in phase voltage for the amplifier. The stereo pickup's two transducers, as in the mono pickup, may be moving coil, moving iron, crystal or ceramic. The 1 mil (0.001) stylus tip of the mono pickup is too coarse for the stereo grooves, which are three mils wide. Some stereo styli are as small as .5 mil. RIAA equalization is standard.

The two transducers must be connected so that their outputs are *in phase* to the speakers. This may require a reversal of the leads to one of the speakers so that their diaphragms move in and out together. Most preamps have a switch for reversing the phase.

Sine wave signals of unequal magnitudes, and great variety in their between-phase relations, create forces that move the point of the cutting stylus in directions that are neither vertical nor lateral but somewhere in between. The stylus may even describe a circle or an ellipse. Moreover, the complex mixture of frequencies from an orchestra creates an almost infinite number of movements of the cutter in both channels, which are repeated by the pickup styli in reproducing the music.

The makers of most commercial phonographs for the mass market have tossed *hi-fi* out the window. *Stereo* has replaced it in their advertising copy. This would seem to reveal a rebirth of conscience on their part, but we're sold on progress in this country, and who is going to object to "moving up" from high fidelity to stereophony?

FM multiplex (stereophonic) broadcasting has recently become available to hi-fi fans throughout the country. The old FM receiver can be retained by adding adapter (cost up to $100) providing it's of high quality.

A few stations equipped with both FM and AM transmitters had been broadcasting stereo for years with one channel on FM and the other on AM, but the relatively poor quality on AM did not make the equipment popular. So why not use twin FM transmitters for the two channels with twin FM receivers? There isn't enough room in the FM frequency spectrum, between 88 and 108 megacycles, for this easy solution. Besides, it would have made obsolete all the old FM sets designed for straight mono reception.

Multiplexing means that a single FM carrier is used for delivering both the left and the right stereo channels. Fundamental to all multiplexing is the use of a *subcarrier*.

A subcarrier is a frequency generated in the transmitter much lower than the carrier. An information frequency modulates the subcarrier before it, in turn, modulates the carrier. Color TV uses a subcarrier. Missiles and satellites transmit a carrier modulated by many subcarriers, each with its own class of information.

The technical demands of a system of FM stereo are very stringent. Compatibility with FM mono is first of all necessary. Low distortion is also a must. Signal separation between the two channels should be at least 30 db (each signal 30 db greater than any leakage from the other channel), and both channels must modulate the carrier wave close to 100 per cent so that the strength of the received signal, both for mono and stereo, is not appreciably reduced. In an effort to meet these demands as many as 17 different methods had been proposed.

After years of study, including a long series of tests, the FCC finally selected a system that is largely a composite of those recommended by General Electric and Zenith. Briefly, here is how it works. The output from the two microphones, left (L) and right (R), is combined (added in phase) and used to frequency-modulate the carrier frequency. (L + R is the formula.) You use two mikes instead of one, as though no stereo were involved. The signal from this modulation is available for the old mono FM sets.

The required subcarrier is only 38,000 cycles (38 kc.). It is amplitude modulated by the difference between the left (L) and the right (R) mike outputs, or L − R. L − R is obtained by combining them in opposite phase in a matrix, as illustrated in Fig. 67.

The amplitude-modulated (by L − R) 38 kc. subcarrier or subchannel then *frequency-modulates* the main carrier. Thus the main carrier is frequency-modulated both by L + R and L − R (and its subcarrier).

If the standard FM receiver is of good quality, it can be used for stereo by adding a multiplex adapter, as shown in Fig. 68. The composite signal, which is the carrier modulated by L + R's and L − R's subcarrier, passes through all the receiver circuits up to audio amplification. These include the r-f amplifier, the mixer-oscillator, which reduces the carrier to the 10.7-mc. intermediate frequency; the i-f amplifiers, the limiter, or limiters, and the discriminator. (The discriminator, you'll recall, is the FM detector that changes a rise in frequency to a current flow in one direction and a fall in frequency to a current flow in the opposite direction). At the discriminator output, the L + R signal is available for mono reception. For stereo, the L + R goes to the matrix, as illustrated, to combine with the L − R signal.

The L − R signal is separated from the discriminator output by means of a bandpass filter. Since it has been amplitude-modulated, it must pass through an AM detector before proceeding to the matrix.

The matrix produces the left channel by adding L + R to L − R. Since the two Rs are 180 degrees out of phase they cancel.

$$L + R$$
$$L - R$$
$$\overline{}$$
$$2L \quad O$$

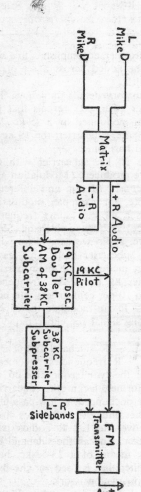

Fig. 68. Block diagram of an adapter for FM multiplex stereo reception.

69. Block diagram for multiplex stereo transmission.

By reversing the L — R leads to the matrix, before adding to the L + R, we obtain the right channel.

$$\begin{array}{r} L + R \\ - L + R \\ \hline O \quad 2R \end{array}$$

Fig. 67 indicates how a transformer accomplishes this sorcery, though the voltages can be added across a set of resistors to obtain the same results.

We have omitted many explanatory details until now. The block diagram of the transmitter (Fig. 69) reveals that the transmitter's carrier wave is also modulated by a 19-kc. frequency. In order to obtain the 38-kc. subcarrier, the 19 kc. is doubled by amplifying its second harmonic.

In amplitude modulating the 38 kc. subcarrier with the audio from L — R, *sidebands* are produced. (Modulation also heterodynes.) These sidebands consist of sum and difference frequencies. A 5 kc. audio note, for example, modulating the 38 kc. subcarrier, produces a 43 kc. (sum) frequency sideband and a 33 kc. (difference) frequency sideband. Since these sidebands contain the 5 kc. audio we can, if we wish, *suppress* (filter out) the 38 kc. subcarrier itself. This leaves the sidebands alone to frequency-modulate the main carrier.

The modulation of the carrier by the 19-kc. frequency, called the pilot carrier, is kept to 10 per cent or less, but both L + R and the sidebands of L — R can modulate the carrier 90 per cent. The result is a strong signal for either mono or stereo reception.

The 23-53 kc. range of the L — R sideband frequencies are so synchronized with the L + R frequencies (by means of the 19-kc. pilot carrier) that the two frequencies do not interfere with each other. The former, lacking its subcarrier, passes through the discriminator unchanged. A low-pass filter (one that passes all the frequencies under 15,000 cycles at the discriminator output) will remove the L + R audios from the sideband frequencies (Fig. 68). And all the sideband frequencies may be recaptured by means of a 23-53-kc. band pass filter, the same kind of a filter that is used for the mid-range speaker in a three-way cross-over network.

When we reinsert between the L — R sidebands the 38,000-cycle subcarrier that was suppressed in the transmitter, the original signal is reconstructed in the receiver. As the subcarrier was amplitude-modulated, a diode detector provides the L — R signal for the matrix.

We can obtain the receiver's 38 kc. by amplifying the second harmonic of the 19-kc. pilot carrier, or by using a 38-kc.

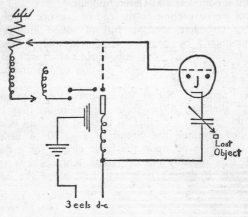

FIG. 70. Sketch of a bio-electronic inversion comparator with a forward feedback, diatomaceous grid amplifier, reversible diode, and paramagnetic pickup. It runs on three small electric eels in series, and is useful for locating things around the house, such as pencils, small change to pay the Good Humor Man, and grandma's glasses.

oscillator, which is frequency-controlled by the 19-kc. frequency. Control is necessary for synchronization.

FM transmitters *pre-emphasize* the audio frequencies above 500 cycles to increase the signal-to-noise ratio. For this reason, these frequencies require *de-emphasis* in the receiver. The de-emphasis circuit may consist merely of a resistor in series with a capacitor across the line. Such a circuit removes more and more of the energy as the frequency rises. I recommend one with a time constant of 75 micro-seconds (T = RC). For example, a 100,000-ohm resistor can be used with a .00075-μfd capacitor.

There are many possible variations of the circuit described here for unscrambling an FM stereo signal. In purchasing an adapter, you should select one from the manufacturer whose name appears on your FM receiver. The audiophile who is without an FM receiver, or is using one with a limited band width (for good multiplex reception, you need a wider band width in the i-f stages and the discriminator than is present in many FM sets) can now obtain a receiver with an integrated stereo adapter. The novice embarking upon a hi-fi career should consider one of the new stereo preamp-amplifiers, with a complete FM stereo built in.

So now we have come to the end of our tale which, if not told by a complete idiot, is full of sound and fury, the latter minimized by negative feedback; and if it signifies nothing to you, check your impedance match with my style, for I have stayed on the linear portion of the characteristic curve as best I could.

Diane Vance and Scotty Bivens drew the diagrams from my rough sketches. Although in the beginning they couldn't tell a cathode from a decibel, at the end they had become so expert they invented the circuit of Fig. 70. A little study reveals they have hit upon the basic circuit for a computer that will translate any one of Nikita Khrushchev's speeches into Swahili. It doesn't help any. Left alone the computer plays solitaire.

Index

Other SIGNET SCIENCE Books

(60¢ each)

UNDER THE SEA WIND by Rachel L. Carson

Life among birds and fish on the shore, in the open sea, and on the sea bottom. (#P2239)

THE SEA AROUND US by Rachel L. Carson

The outstanding bestseller and National Book Award winner, an enthralling account of the ocean, its geography and its inhabitants. (#P2361)

THE WEB OF LIFE by John H. Stoirer

An easy-to-understand introduction to the fascinating science of ecology, showing how all living things are related to each other. (#P2265)

THE WELLSPRINGS OF LIFE by Isaac Asimov

The chemistry of the living cell and its relation to evolution, heredity, growth and development. (#P2066)

MAN: HIS FIRST MILLION YEARS by Ashley Montagu

A vivid, lively account of the origin of man and the development of his cultures. customs, and beliefs. (#P2130)

THE STARS by Irving Adler

A popular summary of the nature, movement, and structure of the stars and a simple explanation of the evidence which has led scientists to the basic conclusions of astronomy. (#P2093)

THE SUN AND ITS FAMILY by Irving Adler

A popular book on astronomy which traces scientific discoveries about the solar system from earliest times to the present. Illustrated. (#P2037)

Other SIGNET SCIENCE Books

Science In Our Lives *by Ritchie Calder*. An exciting factual story of the beginning and development of modern science, the relationship between its special fields—astronomy, chemistry, physics, biology—and its impact upon our daily lives.
(#P2124—60¢)

Magic House of Numbers *by Irving Adler*. Mathematical curiosities, riddles, tricks, and games that teach the basic principles of arithmetic. (#P2117—60¢)

Thinking Machines *by Irving Adler*. How today's amazing electronic brains use logic and algebra to solve a great variety of problems. (#P2065—60¢)

The New Mathematics *by Irving Adler*. The first book to explain—in simple, uncomplicated language—the fundamental concepts of the revolutionary developments in modern mathematics. (#P2099—60¢)

Mathematics In Everyday Things *by William C. Vergara*. In fascinating question and answer form, and illustrated with diagrams, this book shows how the basic principles of mathematics are applied to hundreds of scientific problems.
(#T2098—75¢)

Electronics for Everyone (revised, expanded), *Monroe Upton*. Today's discoveries in the field of electronics, and a forecast of its role in the future. (#T2164—75¢)

Medicine and Man, *Ritchie Calder*. Important medical events and discoveries from earliest times to the present.
(#P2168—60¢)